101 KEY IDEAS

GENETICS

Lorna McPherson

Oct 2000

101 KEY IDEAS

GENETICS

Morton Jenkins

TEACH YOURSELF BOOKS

For UK orders: please contact Bookpoint Ltd, 78 Milton Park, Abingdon, Oxon OX14 4TD. Telephone: (44) 01235 400414, Fax: (44) 01235 400454. Lines are open from 9.00–6.00, Monday to Saturday, with a 24 hour message answering service. Email address: orders@bookpoint.co.uk

For USA & Canada order queries: please contact NTC/Contemporary Publishing, 4255 West Touhy Avenue, Lincolnwood, Illinois 60646–1975, USA. Telephone: (847) 679 5500, Fax: (847) 679 2494.

Long renowned as the authoritative source for self-guided learning – with more than 30 million copies sold worldwide – the *Teach Yourself* series includes over 200 titles in the fields of languages, crafts, hobbies, business and education.

British Library Cataloguing in Publication Data
A catalogue record for this title is available from The British Library.

Library of Congress Catalog Card Number: On file

First published in UK 2000 by Hodder Headline Plc, 338 Euston Road, London, NW1 3BH.

First published in US 2000 by NTC/Contemporary Publishing, 4255 West Touhy Avenue, Lincolnwood (Chicago), Illinois 60646–1975 USA.

The 'Teach Yourself' name and logo are registered trade marks of Hodder & Stoughton Ltd.

Cover design and illustration by Mike Stones.

Typeset by Transet Limited, Coventry, England.
Printed in Great Britain for Hodder & Stoughton Educational, a division of Hodder Headline Plc, 338 Euston Road, London NW1 3BH by Cox & Wyman Ltd, Reading, Berkshire.

| Impression number | 10 9 8 7 6 5 4 3 2 1 |
| Year | 2005 2004 2003 2002 2001 2000 |

Contents

Introduction

Welcome to the **Teach Yourself 101 Key Ideas** series. We hope that you will find both this book and others in the series to be useful, interesting and informative. The purpose of the series is to provide an introduction to a wide range of subjects, in a way that is entertaining and easy to absorb.

Each book contains 101 short accounts of key ideas or terms which are regarded as central to that subject. The accounts are presented in alphabetical order for ease of reference. All of the books in the series are written in order to be meaningful, whether or not you have previous knowledge of the subject. They will be useful to you whether you are a general reader, are on a pre-university course, or have just started at university.

We have designed the series to be a combination of a text book and a dictionary. We felt that many text books are too long for easy reference, while the entries in dictionaries are often too short to provide sufficient detail. The **Teach Yourself 101 Key Ideas** series gives the best of both worlds! Here are books that you do not have to read cover to cover, or in any set order. Dip into them when you need to know the meaning of a term, and you will find a short, but comprehensive account which will be of real help with those essays and assignments. The terms are described in a straightforward way with a careful selection of academic words thrown in for good measure!

So if you need a quick and inexpensive introduction to a subject, **Teach Yourself 101 Key Ideas** is for you. And incidentally, if you have any suggestions about this book or the series, do let us know. It would be great to hear from you.

Best wishes with your studies!

Paul Oliver
Series Editor

Acknowledgement

The author expresses his sincere thanks to Helen Green of Hodder & Stoughton for her help and encouragement throughout the production of this book. It is also a pleasure to thank Dr Sue Noake, Headteacher of Lewis Girls Comprehensive School, Ystrad Mynach, for her help and advice during the preparation of the text.

Achondroplasia

Achondroplasia is a genetic condition in which a person is of very short stature because of restricted growth of bone and cartilage. Some conditions that cause restricted growth affect the whole body equally, but the growth of a person with achondroplasia is disproportionate because growth is most restricted in the long bones of the arms and legs, while the trunk is near to normal size.

The pattern of inheritance for achondroplasia is autosomal dominant. Where one parent has achondroplasia, there is a 50 per cent (one in two) chance of every child being affected. If both parents have achrondroplasia, every child has a 50 per cent (one in two) chance of having achrondroplasia and a 25 per cent (one in four) chance of being unaffected. There is also a 25 per cent chance of the baby inheriting the achondroplasia gene from both parents, in which case it will be severely affected and will die soon after birth.

However, most people with the condition have no family history of achondroplasia. They are probably affected because of a spontaneous genetic mutation which took place in the egg or sperm before fertilization. The condition occurs in all races and in both sexes. About one in every 25,000 babies born in Britain will have achondroplasia.

The average height for an adult with the condition is 1.12–1.45 m. The head is slightly enlarged with a flattened nose and a prominent forehead, and the fingers will be short and stubby. The lower legs may be bowed, and curvature of the lower spine can sometimes cause pain and back problems in later life. Children with achondroplasia may take longer to stand. Ear infections may be more common, leading to loss of hearing if not treated. Because of the shape of the lower jaw, teeth may be crowed and badly aligned. Some people wrongly assume that a small body invariably means a small brain but people with achondroplasia have the same range of intelligence as the general population. Many of the problems people with achondroplasia have to face are to do with how society treats them.

Allopolyploidy

A type of polyploid mutation involving the combination of chromosomes from two or more different species. Allopolyploids usually arise from doubling of chromosomes of a hybrid between two species (this is called amphiploidy). The properties of the hybrid – for example, greater vigour and adaptability – are retained in the allopolyploid in subsequent generations and such organisms are often highly successful.

The most commonly encountered type is the allotetraploid, which has two complete sets of genes from each of the two original parent species. The Russian cytologist Karpechenko (1928) synthesized a new genus from crosses between vegetables belonging to different genera, the radish (*Raphinus*) and the cabbage (*Brassica*). These plants are fairly closely related and belong to the same family (*Cruciferae*) which include mustard. Each has a diploid chromosome number of 18, but the chromosomes in the radish have many genes that do not occur in the cabbage, and vice versa. Karpechenko's hybrid had in each of its cells 18 chromosomes, nine from the radish and nine from the cabbage. Members of the very dissimilar sets of genes failed to pair during meiosis and the hybrid was largely sterile. A few 18-chromosome gametes were formed, however, and a few allotetraploids were therefore produced in the F_2 generation. These were completely fertile, because two sets each of radish and cabbage chromosomes were present and pairing between homologous chromosomes took place at meiosis. The allotetraploid, or amphidiploid, was named *Raphanobrassica*. Unfortunately, the resulting plant has the root of a cabbage and leaves of a radish and so is of no economic value. The method does, however, illustrate the principle of producing fertile interspecific or intergeneric hybrids.

At least half of all naturally occurring polyploids are allopolyploids. Cultivated wheat provides a good example. Cultivated wheat has 42 chromosomes, representing a complete diploid set of 14 chromosomes from each of three ancestral types.

see also...

Meiosis; Mutagens; Polyploidy

Autopolyploidy

A type of polyploid mutation involving the multiplication of chromosome sets from only one species (*auto* = self; *poly* = many; *ploid* = chromosome number). Autopolyploids may arise from the fusion of diploid gametes of the same species that have resulted from chromosomes failing to separate at meiosis. Alternatively, like allopolyploids, they may arise by the failure of chromatids to separate during the mitotic division of a zygote. The hybrid formed as a result of autopolyploidy may be fertile or sterile, depending on the number of chromosome sets. Hybrids with an even number of homologous chromosome sets (4, 6, 8...28) will be fertile because chromosome pairing is possible at meiosis.

When a gamete, unreduced at meiosis, remains diploid (2×) it may fuse with a normal haploid gamete (1×), giving a triploid offspring (3×). The union of two unreduced (2×) gametes gives an autotetraploid. Triploids can also arise from crossing between diploids and tetraploids. Another source is spindle failure at mitosis, which gives direct doubling of the somatic chromosome number and leads to the production of a polyploid cell. Derivatives of the doubled-up cell could be 4×, 6× or 8×, and may them form a polyploid part like a branch in an otherwise 2×, 3× or 4× plant. Autopolyploids have chromosome sets that are all homologous with one another.

Chrysanthemum, the celandine (*Ranunculus ficaria*), the hyacinth (*Hyacinthus orientalis*), and some varieties of Cox's orange pippin (*Malus pumila*) represent this type of polyploidy. Artificially induced autoplyploidy, with the use of colchicine, has been used to produce new and more vigorous varieties of commercial crop plants such as tomatoes and sugar beet.

> **see also...**
>
> *Allopolyploidy; Chromosomes; Meiosis; Mitosis; Polyploidy*

Beadle and Tatum

George Wells Beadle and Edward Lawrie Tatum provided firm evidence of the link between genes and the control of enzyme production. In fact, they proved that one gene is responsible for controlling the production of one enzyme. They carried out their classic investigations using the mould *Neurospora*, which is a pinkish fungus, often found growing on stale bread.

Beadle was born in Nebraska in 1903, where his father ran a small farm. He gained his doctorate at Cornell University in 1931 as a geneticist, having been fascinated by the research carried out by Thomas Hunt Morgan on fruit flies. After several years carrying out research at the *Institut de Biologie Physico-Chimique* in Paris, he returned to the United States. In the autumn of 1937, Beadle was working at Stanford University, where he met Tatum.

They searched for a gene mutation that would influence chemical reactions in living things. They chose *Neurospora* as their research material because its diet consists of sugar, minerals and biotin, one of the vitamin B group. With this simple diet, *Neurospora* can synthesize all of the 20 amino acids and all the vitamins it needs (other than biotin).

Because *Neurospora* can synthesize so many substances, it followed that the fungus must have the genes to control the production of all the enzymes involved. Beadle and Tatum predicted that, by causing some of the mould's genes to mutate, they should be able to produce moulds with errors of metabolism that could be identified chemically. Their analogy was:

> If cars are observed as they emerge from an assembly line, it is not possible to determine what each worker did as part of the cars' manufacture. However, if one could replace able workers one by one with defective ones and then observe the results in the products, it would be possible to determine that one worker put on the radiator, another added the fuel pump, and so on.

Beadle and Tatum gained a share of the 1958 Nobel Prize for medicine and physiology.

see also...

Mutation; Sex-linked genes

Bottle-neck effect

opulations may sometimes be reduced to low numbers through periods of seasonal climatic change, heavy predation, disease or catastrophic change involving volcanic eruptions or other natural disasters. As a result, only a small number of individuals remain in the gene pool to contribute their genes to the next generation. The small sample that survives will often not be representative of the original, larger, gene pool and the resulting allele frequencies may be severely altered. In addition to this 'bottle-neck' effect, the small surviving population is also often subject to inbreeding and genetic drift.

In a nutshell, the life history of a population subjected to a bottle-neck is as follows:

★ A large population exists with very much genetic diversity.
★ The population crashes to a very low number, loses nearly all of its genetic diversity and almost becomes extinct.
★ The population grows to a large size again but has lost much of its genetic diversity.

The principle has been illustrated by studies of the world population of cheetahs and can be summarized as follows.

In recent years, the world population of wild cheetahs has declined to fewer than 20,000. With modern techniques of genetic profiling, DNA analysis has found that the total cheetah population has very little genetic diversity. Cheetahs appear to have narrowly escaped extinction at the end of the last ice age, 10,000–20,000 years ago. Perhaps a single pregnant female survived in a cave and produced a litter. All modern cheetahs seem to have arisen from this one surviving litter – accounting for the lack of diversity. It has been estimated that during the last ice age 75 per cent of all large mammals died out (including mammoths, cave bears and sabre tooth tigers). The lack of genetic variation in cheetahs has led to sperm abnormalities, decreased fecundity, high cub mortality and sensitivity to disease. Since the genetic bottle-neck there has been insufficient time for random mutations and produce new genetic variations.

see also...

Genetic drift; Genetic profiling

Cancer

The suggestion that genes play a role in the development of cancer is not a new idea. In fact, as long ago as 1914, the German zoologist Boveri formulated the 'somatic mutation' hypothesis for the origin of cancer. Boveri's hypothesis suggests that at least some cancers result from somatic mutations (mutations in cells other than sex cells – from *soma* = body) affecting the growth patterns of these cells. Somatic mutations cannot be inherited because they do not affect the sex cells. In addition, single genes, some dominant and some recessive, can bring about conditions conducive to cancer. Also, polygenes with additive effects seem to be responsible for some cases of cancer. Such single gene and polygene conditions can be transmitted from generation to generation.

The first two genes (oncogenes) to be discovered that are linked to breast cancer are *BRCA1* and *BRCA2*. They increase a woman's chance of suffering from breast cancer. About one in 12 women in Britain will develop breast cancer at some stage in their lives and about 34,000 cases are recorded each year. However, only a small proportion of these have genetic causes (about 5–7 per cent).

The pattern of inheritance for genetic breast cancer is autosomal dominant. This means that in families affected by genetic breast cancer each child has a 50 per cent (one in two) chance of inheriting the faulty gene. Women who inherit the *BRCA1* or *BRCA2* gene have an 80–90 per cent chance of developing breast cancer. It is very unusual for men to suffer from breast cancer (although it is not impossible), but men can carry a copy of the faulty gene and pass it on to their children. So genetic breast cancer could occur in a family after a couple of unaffected generations because it has been passed down the male line. *BRCA1* was found in 1994 on chromosome 17 and *BRCA2* was found in 1995 on chromosome 13. Faulty copies of the *BRCA1* gene are believed to be present in about half of the families in Britain with genetic breast cancer. *BRCA2* is seen in about one-third.

Chromosomes

The existence of chromosomes was discovered in 1848 by the self-taught amateur scientist Wilhelm Friedrich Hofmeister, while he studied how plant cells divide. Hofmeister worked in his father's bookselling and publishing business in Leipzig, Germany, but spent all of his spare time at his microscope, studying cells. It is said that he often rose at four o'clock in the morning in the summer so as to get in two hours' work before going to the office. Eventually, after publishing many research papers in the 1840s and 1850s, he was offered an academic post and in 1863 became Professor of Botany at Heidelberg University.

Hofmeister carefully observed that, before a cell divides to produce new cells, its nucleus divides to produce two daughter nuclei. He also recorded that the nucleus first resolves itself into smaller bodies, which could be stained with certain aniline-based dyes. He called these bodies chromosomes, meaning 'coloured bodies' (*chromos* = colour; *soma* = body). Eventually, scientists found that the nucleus of any plant or animal cell, when about to divide, resolves itself into rod-shaped chromosomes. Further, the number of these chromosomes is always the same in each species of animal or plant, although it varies from one species to another. Humans have 46 chromosomes in all of their cells – except the sex cells, where the number is halved because otherwise the normal number of chromosomes (diploid number) would double every time fertilization took place.

A common misconception students have is that the more complex the organism, the more chromosomes it has. This is not the case. For example, the single-celled animal *Amoeba* has 50, while a housefly has 12. Chimpanzees have two more than humans. The lowest known number is two, in a species of parasitic roundworm.

Chromosomes appear in homologous pairs, one set originating from the sex cell of the female parent and the other set originating from the sex cell of the male parent. Chromosomes are made mainly from DNA and protein.

see also...

Deoxyribonucleic acid (DNA); Proteins

Chromosome maps

By the end of the nineteenth century, geneticists knew that genes were to be found on chromosomes but they did not know the positions of genes in relation to one another. As early as 1911, they wondered about the possibility of 'mapping' the chromosomes to show the relative positions of genes.

If genes are located on chromosomes, and if specific alleles (an allele is a form of a particular gene) are precisely exchanged through crossovers in meiosis, then the genes for certain characteristics must lie at specific points along each chromosome. Mapping chromosomes involves a technique that is largely based on information from observation of crossovers.

Between 1912 and 1915, Thomas Hunt Morgan and A. H. Sturtevent hypothesized that if genes were arranged linearly along chromosomes, then those lying closer together would be separated by crossovers less often than those lying further apart. Genes lying closer together would thus have a greater probability of being passed along as a unit. In other words, the percentage of crossovers is proportional to the distance between two genes on a chromosome – so percentage crossover is the number of crossovers between two genes per 100 opportunities in meiosis.

As an example, suppose two characteristics, which we can call A and B, show 26 per cent crossover. We can assign 26 crossover units to the distance between the two genes. Then if some characteristic C turns out in breeding experiments to have 9 per cent crossover with B and 17 per cent crossover with A, it would be located between A and B at a point 9 units from B and 17 units from A. After the information from many such crosses has been compiled, a chromosome map can be made that indicates the positions along the chromosome of the genes that code for certain characteristics.

In summary: the further apart any two genes are on the same chromosome, the greater the incidence of crossing over between them.

see also...
Meiosis

Cloning genes

Gene cloning is a process of making large quantities of desired DNA once it has been isolated from a chromosome. Cloning allows unlimited copies of a gene to be produced for analysis or for the production of a protein. Methods have been developed to insert a DNA fragment of interest (e.g. a piece of human DNA) into the DNA of a vector (carrier), resulting in a recombinant DNA molecule or molecular clone. A vector is a self-replicating DNA molecule (e.g. a plasmid, which is a circular piece of DNA found in bacteria). Viral DNA may also be used and both types of vector are able to transmit a gene from one organism to another.

All vectors have the following characteristics:

★ they are able to replicate inside their host organism
★ they have one or more sites at which a restriction enzyme can cut
★ they have some kind of generic marker that allows them to be easily identified.

Organisms such as bacteria, yeasts and viruses have DNA which behaves in this way. Large quantities of the desired gene can be obtained if the recombinant DNA is allowed to replicate in an appropriate host. Plasmid vestors, found in bacteria, are prepared for cloning as follows:

1 A gene of interest is isolated from human cells.
2 An appropriate plasmid vector is isolated from bacteria.
3 Human DNA and plasmid are treated with the same restriction enzyme to produce identical 'sticky ends' at the points of cutting.
4 Restriction enzymes cut the plasmid DNA at its single recognition sequence of bases, disrupting the antibiotic (tetracycline) resistance gene.
5 The DNAs are mixed together and the enzyme DNA ligase added to bond the sticky ends.
6 Recombinant plasmid is introduced into a bacterium by adding the DNA to a bacterial culture where some bacteria take up the plasmid from the culture.

see also...

Gene splicing; Protein synthesis; Recombinant DNA

Cloning whole animals

The first pioneers in the field of animal cloning used frogs and carried out their research in the 1970s. A team led by John Gurdon at the University of Cambridge transplanted nuclei from the skin cells of adult frogs into frog eggs which lacked nuclei. Some embryos grew into tadpoles but none metamorphosed into frogs.

Early in 1997 genetics entered a new era, with the arrival of Dolly the sheep. As a product of cloning rather than of fertilization, she was the most remarkable mammal ever to have been born – probably the most famous sheep since the one that provided Jason of the Argonauts with its fleece in ancient mythology. Dolly was created from a cell taken from the udder of a six-year-old ewe. Proving that scientists have a sense of humour (albeit sexist), she was called Dolly after a world-famous Country and Western singer who happened to be well blessed with mammary gland tissue.

The cloning of sheep from different cells of the same embryo had been successful in 1996, but the difference with Dolly was that all her DNA came from one cell of an adult sheep.

The work was done at the Roslin Institute in Edinburgh, and Ian Wilmut of the Roslin team recorded that the process had a success rate of one in 277 attempts. If it could be applied to humans, it could mean that each one of us could have clones made from our own tissue, containing the same DNA.

The stages in the process may be summarized thus.

1 An udder cell is taken from the donor sheep.
2 An egg cell is taken from another sheep and its nucleus (with its DNA) removed.
3 The empty egg cells are fused with the udder cells by subjecting them to an electric current.
4 The resulting dividing cells are cultured for a week until an early embryo forms.
5 This early embryo (a blastocyst) is implanted into a surrogate ewe's womb.

In 1997 Australian researchers cloned 470 genetically identical cow embryos from a single embryo.

Cloning whole plants

A clone is a group of genetically identical organisms derived from a single parent. In plants, micropropagation is a method of cloning. The basis of this is that any differentiated plant cell has the potential to give rise to all the different cells of an adult plant if cultured in the correct media. Micropropagation is widely used for the rapid multiplication of commerically important plant species with genotypes that express desirable characters, as well as in the recovery programmes for endanged species. Forest productivity can be improved as well as wood quality, resistance to disease, pollutants and insects. Factors that affect the success of micropropagation via tissue culture include:

★ quality of the plant material (e.g. buds) being cultured
★ composition of the culture media
★ plant hormone levels
★ lighting and temperature.

The stages in cloning using tissue culture are:

1 Stock plants as free from pests and pathogens as possible are selected.

2 Small pieces are cut from the plant (explants). These may be flower buds, leaves, stem tissue with nodes (leaf buds) or small sections of growing tissues from stem tips.

3 The surfaces of the explants are sterilized using sodium hypochlorite.

4 Under sterile conditions the explants are transferred to a culture vessel. The culture medium contains plant hormones, which control growth.

5 Cultures are incubated for 3–9 weeks at 15–30°C with 10–14 hours of light per day.

6 An undifferentiated mass of cells (called a callus) develops.

7 New shoots that develop are removed from the explant and placed on a new culture medium. The process is repeated every few weeks so that a few plants can give rise to millions of plants.

8 The culture plants are acclimatized in special greenhouses before they are planted outside.

9 If the callus is suspended in a liquid nutrient medium and broken up mechanically into individual cells, it forms a plant cell culture that can be maintained indefinitely.

Co-dominance

When two pure-breeding parents with contrasted characteristics are crossed to produce the F_1 generation, the F_1 is heterozygous – it has both alleles for the characteristics. Each sex cell of the F_1 organism will contain the allele for only one of the characteristics. When these sex cells combine to form the F_2 generation, you would expect to see a ratio of one homozygous dominant:two heterozygous:one homozygous recessive – this is the basis of Mendel's Law of Segregation. However, the principle relies on there being characteristics which are either dominant or recessive – but some characteristics are neither dominant nor recessive and are therefore co-dominant or incompletely dominant. In these cases, the F_1 will show a blending of characteristics and the expression of genes in the F_2 will not produce the classic Mendelian 3:1 ratio.

For example, petal colour is often determined by genes that show absence of dominance. In some species, such as the morning glory flower, if a red-flowered plant is crossed with a white-flowered plant, the F_1 generation is entirely pink-flowered plants. When these are self-pollinated, the F_2 generation comprises the following ratios: one pure red-flowered plant:two pink-flowered plants:one pure white-flowered plant. This type of inheritance is due to incomplete dominance because an allele controls the production of an enzyme which regulates the development of the red pigment. Another allele, expressed as white, does not allow the production of the enzyme. If a plant has two allels for the enzyme, it produces red flowers; if it has both the red and white alleles, it produces pink flowers; if it has two white alleles, it produces white flowers.

Co-dominance is also seen in the colour of the coats of short-horn cattle. In a cross between a red bull and a white cow, the calves are intermediate in colour and are called roan. Their coats contain a mixture of red and white hairs. When two roans are mated the calves will be in the ratio of one red:two roans:one white.

see also...
Mendelism

Continuous variation

This is a form of quantitive variation in which a character has an average value and members of the same species or population show various gradations from this value.

A typical example is height in humans. It is produced by polygenic inheritance, where a single trait is controlled by multiple genes. Because of polygenic inheritance for height, people are *not* either tall or short; instead, there is a gradual variation that enables us to generate a smooth, gradual, and continuous bell-shaped curve as each gene adds its own small influence. It can thus be assumed that if some genes contribute to tallness and others produce shortness, taller individuals are likely to possess a disproportionate number of the 'tall' genes. On the other hand, those with a disproportionate number of 'short' genes will be short. By the same token, this is why parents of average height can produce exceptionally tall or short offspring. Imagine that there are only seven genes that influence height (T for tall and S for short). If parents are STTTSSS × STTTSST, they might have offspring that are TTTTTTS (very tall), SSSTSSS (very short), TSTSTST (intermediate), or any number of other possible combinations and heights.

The principle can be illustrated in the following figure.

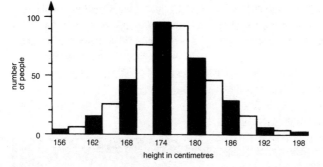

Height in humans shows continuous variation – many different heights are possible

Cystic fibrosis

Cystic fibrosis is an inherited condition which affects mainly the lungs and the pancreas. The disorder is characterized by the inability to pump salt in and out of cells. Thick, sticky mucus builds up in the bronchioles of the lungs, allowing bacteria to collect and causing infections. The digestion of food becomes difficult because of blockage of ducts in the pancreas.

The pattern of inheritance for cystic fibrosis is autosomal recessive. A person who inherits one faulty allele for cystic fibrosis will be a carrier. Carriers are not affected but can pass the faulty allele on to any children they may have. If one or both parents is a carrier, there is a 50 per cent (one in two) chance that each child will also be a carrier. A child who inherits two copies of the faulty allele (one from each parent) will have cystic fibrosis. If both parents are carriers, there is 25 per cent (one in four) chance of this happening.

Cystic fibrosis is Britain's most common life-threatening inherited condition. Every week five babies are born with the condition. It is estimated that approximately one person in 25 in Britain is an unaffected carrier (about 2 million people or 4 per cent of the population).

The search for the defective gene on human chromosomes began in 1987, and the gene was finally tracked down by Lap Chee Tsui, working at the University of Toronto in 1990. By 1993, screening for the gene became routine. Analysis of amniotic fluid identifies fetuses who are afflicted by cystic fibrosis at 18 weeks with a 95 per cent success rate.

There is no cure for cystic fibrosis but treatments are improving and people with it are living longer into adult life. Gene therapy in which working copies of normal dominant genes are introduced into the lungs of sufferers holds hopes for the future. Heart and lung transplants have also improved the quality of life for some people with cystic fibrosis.

see also...
Gene therapy

Deletion mutation

Sometimes block mutations cause genetic imbalances which usually disrupt the development of an organism. Some can result from errors in the crossing over process in meiosis; others may result from the action of mutagens. One form of such a mutation is known as deletion, where pieces of chromosomes are lost.

★ A break occurs at two points on the chromosome and the middle piece falls out.
★ The two ends then rejoin to form a chromosome deficient in some genes.
★ Alternatively, the end of a chromosome in the region of the centromere is lost.

If the part of a chromosome in the region of the centromere is lost by deletion, it is like a car without a driver: it has nothing to steer it in the proper direction in mitosis or meiosis. Hence, the chromosome will not reach the poles of the spindle and will not be included within the nucleus that is formed. Left out in the cytoplasm, it will soon be broken down into its component parts by cellular enzymes and the genes it contained will be lost in the cell. As

there are two of each kind of chromosome, however, alleles of lost genes will be present on the homologous chromosome, but the cell will be haploid for these genes. A large deletion will upset the balance of genes and cause very harmful, or even lethal, phenotypic effects. A very small deletion may not cause any serious upset in the gene balance, but it can allow recessive genes on the homologous chromosome to be expressed. A person with one blue eye and one brown eye, or a person who has a blue sector in an otherwise brown eye, can be produced when a small deletion in a chromosome in a somatic cell in early embryonic development removes the dominant allele for brown pigmentation. One eye may have descended from the cell with the deletion and will be blue, but the other eye will have been formed from cells without the deletion and will be brown.

see also...

Duplication mutation; Gene mutation; Inversion mutation; Lethal alleles; Meiosis; Mutosis; Mutagens; Mutation; Translocation mutation

Deoxyribonucleic acid (DNA)

The search for the structure of the hereditary material is one of the most interesting stories in modern science. By the 1950s, the role of chromosomes in inheritance was well established. The intriguing questions were: What are they? What are they made of? Chemical analysis showed that the chromosomes themselves were composed of protein and an acid called deoxyribonucleic acid, or DNA (a nucleic acid containing the sugar deoxyribose). By 1950 it was generally accepted that the hereditary material itself was the DNA component of the chromosomes. But how was the DNA arranged in the chromosome? What did it look like? How did it duplicate itself during cell division? And, in particular, how could it direct protein synthesis? By the middle of the twentieth century, several intensive research efforts were well under way in the USA and UK. The prevailing notion was that there were three chains in each DNA molecule. Linus Pauling, a well respected American chemist, and Rosalind Franklin of England both favoured this idea, but they disagreed over such details as whether the nitrogen bases stuck outward from the molecule or inward. Then a brilliant and confident young American postdoctoral fellow met an ebullient and talkative English graduate student in biophysics at Cambridge University. The results of what transpired were later described by the American, James Watson, in his controversial best-seller *The Double Helix*.

The partnership, it turned out, was a most fortunate one. The Englishman, Francis Crick, had noted that when the DNA portion of the chromosome was photographed by X-ray diffraction techniques a peculiar image appeared – one which could be produced only by molecules arranged in a helical, or spiral, pattern. In 1953, Watson and Crick announced that DNA is made of two chains coiled around one another in the form of a double helix. (You get the idea if you can imagine a ladder twisted so that its rungs remains perpendicular to its sides.)

see also...

DNA double helix; Nucleic acids; Watson and Crick

DNA *double helix*

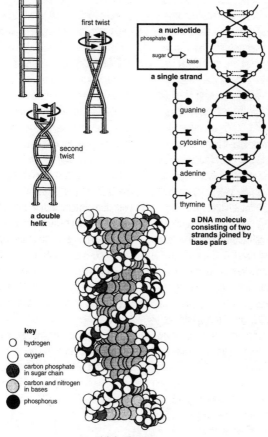

first twist

a nucleotide
phosphate
sugar
base

a single strand

guanine

cytosine

adenine

thymine

second twist

a double helix

a DNA molecule consisting of two strands joined by base pairs

key
○ hydrogen
◯ oxygen
● carbon phosphate in sugar chain
◍ carbon and nitrogen in bases
● phosphorus

a model showing the atoms in a DNA molecule

The Watson–Crick model of DNA

DNA structure

According to the model made by Watson and Crick, the two longitudinal sides of the ladder are composed of alternating sugar and phosphate molecules. The 'rungs' of the ladder are formed by four bases, each containing nitrogen – the purines (bases with two nitrogen rings) adenine (A) and guanine (G); and the pyrimidines (bases with one nitrogen ring) thymine (T) and cytosine (C). These bases, with their sugar and phosphate components, are nucleotides. Watson and Crick described a chain, with one base per sugar–phosphate unit with two bases joining together to form each 'rung'. The bases extend inward from the ladder's sides, they said, and are held together by weak hydrogen bonds. The purines, they pointed out, could not lie adjacent to each other because they are so large that they would overlap – they simply could not fit. The two pyrimidines, with their single nitrogen-containing rings, were too short to reach across. The arrangement could work *only* if a pyrimidine joined with a purine to form the rungs (T to A and C to G).

Furthermore, according to Watson and Crick's deductions, the nucleotides along one of the strands of the double helix could follow any sequence. For example, one might "read" along a strand and find ATTCGTAACGCGT in one segment and something quite different along another segment of the same strand. Also, the strands were very long, so that the necessary complexity for life could be provided by the possible variations in the sequence and the great numbers of nucleotides in any chain. In fact, if the DNA from a single cell were laid out in a straight line, it would be about two metres long! Furthermore, there are about ten thousand million nucleotide parts in the 46 chromosomes of a human cell, so the number of possible variations is mind-boggling.

If the sequence of one strand of DNA is known, then the other can be predicted. The strand in the previous paragraph would have the partner strand TAAGCATTGCGCA.

---

see also...

DNA double helix; Nucleic acids

Dihybrid inheritance

This involves crosses between individuals with two pairs of contrasted characteristics or traits. In 1859, Gregor Mendel carried out experiments involving dihybrid inheritance to check whether traits were inherited independently of each other. He used pea plants that always gave round yellow seeds and crossed them with others that always yielded wrinkled green seeds. If these four traits could be shuffled around quite independently of each other, he would get every possible combination in the first hybrid generation. If these F_1 plants were self-pollinated he obtained a ratio of nine round yellow to three round green to three wrinkled yellow to one wrinkled green. His results are illustrated below.

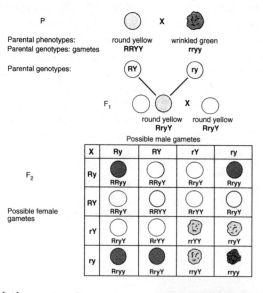

A dihybrid cross using a Punnet square

see also...

Mendelism; Mendel, Gregor; Punnet squares

Discontinuous variation

Discontinous variation is a form of qualitative variation in which a character has two or more distinct forms. It generally occurs when there are two or more alleles of a major gene in a population. In contrast to continuous variation, characters that are coded for by single dominant or recessive alleles produce individuals in distinct groups with no overlapping. Examples are blood groups in humans and the contrasted characters in peas that were the basis of Gregor Mendel's original experiments.

Relatively few traits in humans are known to be controlled by single genes, but two such traits are the ability to roll one's tongue and the pesence or absence of ear lobes. The following digram represents discontinuous variation in pea plants.

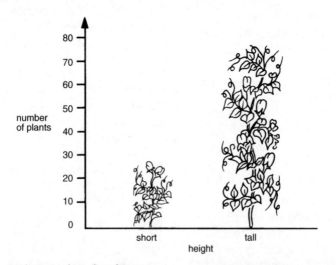

Height in pea plants show discontinuous variation – all plants are either tall or short

see also...

Continuous variation; Mendelism

Dominance

In 1856 Gregor Mendel formulated the principle of dominance. To begin with, Mendel based his conclusions on a carefully planned series of experiments and, more significantly, on a statisical analysis of his results. Mendel planned his investigations using the garden pea *Pisum sativum*. He studied seven pairs of contrasted characters:

1 Seed form – round or wrinkled.
2 Colour of seed contents – yellow or green.
3 Colour of seed coat – white or grey.
4 Colour of unripe seed pods – green or yellow.
5 Shape of ripe seed pods – inflated or constricted between seeds.
6 Length of stem – short 9–18 in) or long (6–8 ft).
7 Position of flowers – axial (along the stem) or terminal (at the end of the stem).

Mendal crossed two true-breeding strains that differed in only one characteristic, such as seed colour. Peas naturally self-fertilize, so for this cross it was necessary to transfer pollen by hand (using a small brush). Mendel called this original parental generation P_1 and designated their first generation offspring the F_1 ('first-filial') generation. When the F_1 plants were allowed to self-fertilize, so that they crossed with each other at random, the offspring resulting from this cross were called the F_2 generation, and so on.

When Mendel crossed his original P_1 plants he found that the characteristics of the two plants did not blend. For example, pure-bred yellow-seeded plants, when crossed with pure-bred green-seeded plants, always produced yellow-seeded plants. Mendel termed the trait that appeared in the F_1 generation the dominant trait. The opposite trait was termed recessive because it appeared to recede into the F_1 but reappear in the F_2. The principle of dominance states that:

Dominance is the expression of only one allele (one alternative form of a gene) when two are present. Recessive is defined as the lack of expression by an allele when a dominant is present.

see also...

Mendelism; Mendel, Gregor

Down's syndrome

A person with Down's syndrome is born with an extra chromosome in each cell (47 instead of 46). The word 'syndrome' describes a collection of features or characteristics which tend to be evident in people with the condition. Down's syndrome is named after doctor John Langdon Down, who first described the condition in 1866. On average, three babies are born with Down's syndrome each day in the UK (about one per 1000 babies). Many people with Down's syndrome enjoy a healthy life and now live to 50–70 years of age.

People with Down's syndrome have some learning disability and special educational needs. It is fairly common for people with Down's syndrome to be born with a heart defect but medical technology can often overcome this problem.

Down's syndrome is not usually inherited: most people who have a baby with Down's syndrome do not have Down's syndrome themselves. At some point when the egg or sperm was made, or perhaps at fertilization, an unusual cell division took place which resulted in an extra copy of chromosome number 21 in all the baby's cells. There is no reason known for certain why this happens: it happens by chance and can happen to anyone, but the chance of it happening increases with the age of the mother.

The chances of having a baby with Down's syndrome are one in 1400 if the mother is 25 years old, rising to one in 110 if the mother is 40 years old, and one in 30 if she is 45 years old. However, most babies with Down's syndrome are born to younger women because the overall birth rate is higher in this age group.

A few people are born with Down's syndrome because they have inherited a chromosome abnormality called translocation from one of their parents. A baby born to a mother with Down's syndrome would also have a high chance of having Down's syndrome, particularly if the baby's father also had the syndrome.

see also...

Non-disjunction; Translocation mutation

Duplication mutation

Duplication is a form of chromosomal block mutation where a part of a chromosome may be duplicated and occur either twice on the same chromosome or on two different non-homologous chromosomes. The two most common features of deletions are:

★ A segment is lost from one chromosome and is added to its homologue. The chromosome with the duplication will become incorporated into a gamete, which may later help to produce an embryo.

★ A segment is taken from its homologue and inserted to produce double copies of some genes. Some genes may be disrupted by this process.

Duplications can arise from errors in crossing over in meiosis. They affect the phenotype because of the altered gene dosage. The effect depends upon the particular segment of a chromosome concerned, but clearly duplications are less harmful than deletions as there is no loss of genetic material.

Duplications are important in evolution. When more than one copy of a gene is present, the redundant one if free to mutate and to evolve a new function.

Duplications are useful in studying the quantitative effects of genes normally present only in pairs in diploid cells. The first duplication to receive critical study was the 'bar eye' variant in the fruit fly, *Drosophila*. The wild type is essentially oval in shape; the bar eye phenotype is characterized by a narrower, oblong, bar-shaped eye with fewer facets.

The classical studies of Calvin B. Bridges in 1936 showed this trait to be associated with the duplication of a segment of the X chromosome, as observed in 'giant' chromosomes of *Drosophila*'s salivary glands. Although deletions are more important than duplications in terms of survival, survival decreases with increasing size of duplication.

see also...

Deletion mutation; Gene mutation; Inversion mutation; Lethal alleles; Meiosis; Mutation; Sickle-cell anaemia; Translocation mutation

Enzymes

Enzymes are large, complex protein molecules with special shapes. They are used in all living things to speed up chemical reactions. The shapes of their molecules are critical to their functions. An enzyme's twisted and coiled molecular stucture forms a kind of groove or slot on its surface. This slot is called the active site because of its role in the enzyme's chemical behaviour. Lining the active site are certain functional groups (side groups with very specific chemical characteristics), which are able to interact with certain combinations of molecules with which they come into contact and to change them in some way. The molecules they act upon are called substrates. The molecular combination resulting from the enzyme's activity is called the product.

Most enzymes are believed to work by placing stress on the arrangement of the substrate molecules, weakening their bonds and allowing them to interact with other molecules in new ways. An inexact fit between enzyme and substrate molecules causes bond changes in the substrate molecule. When the substrate fits into the active site it is something like a key fitting into a lock and has led to the 'lock-and-key' theory of enzyme action. However, this may be an over-simplification because other factors come into play in order to make the 'key' fit the 'lock'. When the interaction between enzyme and substrate is complete, the molecule is called the enzyme–substrate complex. The product then breaks away from the enzyme, which remains unchanged and is ready to work again. The whole process may take place so quickly that a single enzyme may go through several million such cycles each minute in an amazing display of efficiency.

Enzymes are able to bring about several kinds of reactions. For example, they may encourage substrate molecules to join to other molecules – we see this as our cells build large complex molecules from smaller ones. Enzymes can also help to break large molecules into smaller ones. In relation to genetics, enzymes illustrate the importance of genes because genes control the production of proteins and all enzymes are proteins.

Eugenics

Eugenics is the theory that the human race could be improved by controlled selective breeding using individuals with desirable characteristics. Apart from the ethical considerations of whether this is desirable, there are practical problems in the judgement of what are 'desirable' characteristics and with the problem of assessing the role of genetic and environmental influences on development. The earliest ideas of 'improving' society by selective breeding began with Plato in 427 BC. He suggested a system of breeding festivals for gifted men and women, and state-controlled nurseries for the offspring. In Aristotle's *Politics* (383 BC), abortion by choice is advocated along with humane killing of the mentally and physically handicapped.

The term eugenics was coined in 1883 by Francis Galton (1822–1911), a first cousin of Charles Darwin. In his book *Hereditary Genius* (1869) he put forward the idea of judicious marriages between men of genius and women of wealth! Before his death, he endowed a research post in eugenics at University College London, and founded the Eugenics Society. Probably the best-known scientist to take up the Galton research post was Professor Karl Pearson. The extreme racial views of inheritance which spread throughout Europe and the USA in the first half of the twentieth century probably stemmed from his work on eugenics. Scientists with preconceived ideas began falsifying data and many insinuations regarding the inheritance of intelligence and criminality were erroneously made. Eugenics was overtly practised in the USA in the 1920s and 1930s. By 1931, most states had sexual sterilization laws aimed at eliminating the mentally retarded, epileptics and sexual deviants. Over 1000 people were sterilized under these laws in California. At the same time, the belief in the superiority of the Aryan race was growing among fanatics in Nazi Germany, leading to the horrific holocaust.

As recently as 1995 the Chinese government passed a controversial eugenics law which requires people who want to marry to undergo screening for 'inappropriate' genes. Marriages are allowed only if couples with such genes agree to sterilization or long-term contraception.

Fertilization

The starting point for a new, sexually reproducing, organism is the fusion of a single male sex cell with a female sex cell. In this way, the total chromosome number (diploid number) is made of half (haploid number) from the male parent and half from the female parent. The human male sex cell is the sperm, which is too small to be seen without a microscope. The sperm was first described by the Dutchman Antoni van Leeuwenhoek (1632–1723) as a tadpole-shaped cell, so small that millions could fit onto a pin head. In comparison, the female sex cell (or egg) is gigantic, and just visible to the naked eye. The egg was first described by a fellow Dutchman, Regnier de Graaf, when he made microscopic observations of ovaries from small mammals.

Even though sex cells were observed by these pioneer microscopists, scientists were not convinced that they fused during fertilization until the nineteenth century. However, the first investigation, a fascinating experiment in 1780, was carried out by an Italian abbot, Lazzaro Spallanzani (1729–99). He was convinced that the egg carried an already formed embryo. He collected frogs and toads from the river Po in Tuscany at mating time, and fitted the males with little pairs of silk pants! As a result, he found that the eggs produced by the females remained unfertilized. He then filtered the seminal fluid shed by the males through filter paper, removing the sperm from the fluid, and mixed the filtrate with the eggs. No development of the egg took place. In this way, Spallanzani had demonstrated that sperm and egg are needed for fertilization.

Even when the basic idea of joining together the egg and sperm had been accepted, some scientists thought that the job of the sperm was merely to enter the egg and stimulate it to divide. However, in 1875 Oscar Hertwig in Munich, Germany, observed the nucleus of a sea urchin's egg fusing with that of a sperm and recorded that only one sperm reached its goal. Two years later, the same process was described in plants by Eduard Strasburger.

see also...

Chromosomes

Forensic genetics

In 1987 the first case using DNA profiling proved that a suspect who had confessed to a rape and murder was, in fact, innocent. Two girls had been raped and murdered in the same part of the country, but the crimes were committed three years apart. Investigators suspected a connection between the two and soon a suspect was being cross-examined. He admitted to one of the crimes but denied being involved with the other.

Forensic scientists used DNA profiling which showed clearly that both crimes were committed by the same person, but the suspect was not that person. All men from the area were screened and genetic profiles made of 5000 samples. Finally one sample matched those from both crime scenes and the culprit was arrested and convicted. There is now a computerized datebase holding the DNA profiles of all convicted criminals.

In the 1990s a more sophisticated technique for DNA profiling was developed. The new technology required only very small traces of DNA and the reactions in the process happened quickly, reducing the time taken for analysis. The method simulates the replication of chromosomes, producing millions of copies of DNA in a test tube. It is called DNA amplification or the polymerase chain reaction (PCR). The technique detects much smaller minisatellites than those involved in earlier attempts, and the pieces of DNA required are almost 100 times smaller than those needed previously.

The forensic use of DNA profiling is not limited to criminal investigations. The method was used in 1991 when a 5000-year-old frozen body, 'Otzi the iceman', was found in a glacier on the Austrian-Italian alpine border. Some believed that it had been brought there as a hoax from South America but DNA profiling showed that he was more closely related to Northern European races than to inhabitants of South America.

see also...

Cloning whole animals; Genetic profiling

Galactosaemia

Galactosaemia is quite a rare inherited metabolic condition, in which the body cannot convert galactose into glucose. Galactose is the sugar we usually get from lactose, the major sugar in milk and milk products. If the body cannot use it, galactose builds up in the blood and tissues and is converted into harmful by-products which can cause severe damage. The pattern of inheritance for galactosaemia is autosomal recessive. A person who inherits one faulty allele for galactosaemia will be a carrier. Carriers are usually unaffected but can pass the faulty allele on to any children they may have. If one or both parents is a carrier, there is a 50 per cent (one in two) chance that each child will also be a carrier. A child who inherits two alleles for the condition (one from each parent) will suffer from galactosaemia. If both parents are carriers, there is a 25 per cent chance (one in four) of this happening.

The problem arises after galactose is absorbed from the intestine and passes to the liver. Normally the galactose is changed, with the help of an enzyme, to galactose phosphate. This in turn is changed by another liver enzyme, galactose phosphate unridyl transferase (GPT) to glucose phosphate. People who suffer from galactosaemia fail to produce GPT. A fetus that lacks GPT can function normally because its mother's liver handles the galactose for it, but once the umbilicial cord is cut the baby becomes dependent on its own enzymes.

Early diagnosis (within the first two weeks after birth) is vital or problems will arise. These include damage to the liver and to the ovaries and the formation of cataracts in the eyes. If diagnosis is delayed for any length of time the baby will die from liver damage or infection. A person with galactosaemia will need to eat a galactose-free diet for life and so has to avoid all milk and dairy products (lactose is also found in many processed foods and is used in tablets and capsules). Some people with galactosaemia may also have mild learning difficulties.

Gene

William Bateson, an English scientist, read about the discovery of Gregor Mendel's work while travelling by train from Cambridge to London. At once his imagination was fired. For the rest of his life, Bateson was a staunch supporter of Mendel's ideas and he interpreted them and taught them. In 1907 he coined the name 'genetics' for the whole of the study of heredity. Curiously enough, however, Bateson never wholly accepted Mendel's findings. He was enthusiastic about the fact that individual traits were inherited and segregated from each other in the sex cells, but he was uneasy about the idea that particles of matter could carry these on from one generation to another. 'Genetic factors are definite things,' he wrote in 1914, 'either present or absent from any germ cell.' Yet he had also said, a paragraph earlier, that it is unlikley that the 'genetic factors are in any simple or literal sense, material particles.' Mendel had called them 'germinal units'. Bateson could not accept the idea that the inherited characteristics of himself and other people might stem from nothing more than a collection of molecules.

A Danish botanist, Wilhelm Johannsen, was equally uncertain. Although he decided that definite units in the sex cells must determine the herditary characteristics, and christened them 'genes', he did not want to allocate any material basis to them.

It is now known that genes are composed of DNA. They are the smallest units of hereditary material located on chromosomes and determine characteristics of organisms. Genes may exist in a number of forms, called alleles. In a normal diploid cell only two alleles can be present together, one of each of a pair of homologous chromosomes. The alleles may both be of the same type, or they may be different. The segregation of alleles at meiosis and their dominance relationships are responsible for the particulate nature of genes. Genes can occasionally undergo changes, called mutations, to form new alleles.

see also...

DNA structure; Dominance; Gene mutation; Mendelism; Meiosis

Gene interaction

It is possible for genes to interact and provide results which do not conform to the normal Mendelian pattern. Such a case occurs in the inheritance of comb shape in chickens. This was first observed and interpreted by William Bateson and Reginald C. Punnett at the beginning of the twentieth century. Two different forms of comb are called pea and rose. Both of these are true breeding but when crossed together produce a hybrid with pea and rose shape blended together as a different shape, walnut. If two walnut types are mated, the next generation is composed of the following ratios: nine walnuts, three rose, three pea, one single (an entirely new shape).

Bateson and Punnet suggested that there are two independent genes: **P** (dominant) for presence of pea and **p** (recessive), for absence of pea. Also there is **R** (dominant) for presence of rose, and **r** (recessive) for absece of rose. If they appear together, they interact and produce walnut (**RP**). If (**prpr**) occurs there is absence of both genes and the product is the new, single, comb.

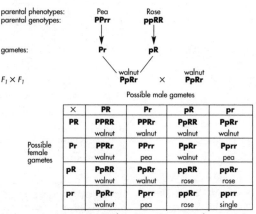

The interaction of two genes controlling comb shapes

Gene mutation

Gene mutations are sometimes called point mutations and are genetic changes affecting the base sequence of a single gene. They may result in the formation of a new allele. A new sequence of bases in DNA will result in a new sequence of amino acids making up a protein. There are two major types of gene or point mutation:

1 Miss-sense substitution, where a single base is substituted by another. This usually results in a new amino acid coded for in the polypeptide chain.
2 Non-sense substitution, where a single base is substituted by another. This results in a new triplet that does not code for an amino acid. The resulting triplet may be an instruction to terminate the synthesis of the polypeptide chain.

Point mutations may be harmful, beneficial or have no effect.

Harmful mutations: There are many examples of harmful mutations that result from alterations to the base sequence in DNA. Examples include:

★ sickle-cell anaemia
★ cystic fibrosis
★ thalassaemia.

Neutral mutations:

★ They may have little or no effect on the survival of an organism or its ability to reproduce.
★ They may result from a 'same sense' mutation, in which the change in the third base sequence still codes for the same amino acid.

Beneficial mutations:
Examples include:

★ bacterial resistance to antibiotics
★ insecticide (e.g. DDT) resistance in insect pests
★ rapid mutation rates in the protein coats of viruses.

see also...

Cystic fibrosis; Deletion mutation; Duplication mutation; Inversion mutation; Lethal alleles; Mendelism; Mutation; Protein synthesis; Sickle-cell anaemia; Thallassaemia; Translocation mutation

Gene splicing

f an organism produces a particular protein, then the organism must have the gene that codes for the protein's production. The skill comes in searching through the organism's chromosomes and finding exactly where the gene is positioned so that it can be cut out and replicated.

One approach is to treat the cells containing the gene with restriction enzymes. These will chop up the chromosomes, cutting them in specific places and leaving bits that can be inserted into the DNA of the host chromosomes. In this way a collection of cells is produced, some of which will carry the gene of interest in their chromosomes. The geneticist must isolate those cells that contain the required gene. This can be done by identifying the cells that make the protein – some proteins interact with certain chemicals (called antibodies), and if such a reaction is detected the gene must be present.

When restriction enzymes act on strands of DNA, they cut the strands at very specific places. The type made by the bacterium *Escherichia*

coli recognizes the following sequence of DNA bases:
     GAATTC
and its complementary sequence:
     CTTAAG
The enzyme breaks the DNA wherever these sequences appear, in the following position:
     G   AATTC
     CTTAA   G
The result is two DNA strands with uneven 'sticky' ends: that is, they tend to stick to other molecules with a complementary sequence of bases. Restriction enzymes do not distinguish the source of DNA – they will cut the specific sequence of bases wherever it exists, and whenever complementary sticky ends meet, they join. After disrupting the source cell and subjecting the chromosomes to a specific restriction enzyme, the genetic engineer is left with pieces of source DNA with specific sticky end containing a particular sequence of nucleotides. The next step is to produce recombinant DNA in the host cell.

see also...

DNA structure; Genetic engineering; Recombinant DNA

Gene structure

The definition of a gene is based on it being a particulate structure responsible for heredity. The definition is based on the observation of the way in which characters are inherited when organisms reproduce sexually. The substance which makes up the chromosomes and which carries information in the genes is DNA. A structural definition of a gene is as follows:

> A gene is a sequence of nucleotide pairs along a DNA molecule which codes for an RNA or polypeptide product.

Both strands of the double helix make up the gene. Only one strand, the transcribed coding strand, contains the information which is directly used for RNA or amino acid assembly. The other strand (the non-transcribed one) contains nucleotide sequences that are complementary to those of the coding strand and is used for replication. The transcribed coding strand is not necessarily the same strand in all of the genes along a chromosome: it switches from one to the other.

Genes that code for polypeptides fall into two main categories:

1 Structural genes, which code for functional proteins (enzymes, hormones, antibodies, etc.)
2 Regulatory genes, which control the activity of other genes.

Polypeptide chains vary in length, and so do the genes that code for them. The average size of a polypeptide chain is 333 amino acids, so the average size of a structural gene is of the order of 1000 nucleotide pairs of DNA.

Transfer (tRNA) and ribosomal RNA (rRNA) moleules are also coded for by genes but they are not synthesized in the same way as proteins. The tRNA and rRNAs are made directly by transcription from the DNA, in the same way as mRNA. Transfer RNA molecules are about 80 nucleotides long, and the genes that code for them are therefore of corresponding length in terms of DNA.

see also...

Gene; Protein synthesis – messenger RNA; Protein synthesis – transcription

Gene switches

The understanding of the genetic code by 1966 marked an end of an era in molecular biology and genetics. Perhaps one of the most important unsolved problems left was to find out what tells genes to work or not to work. It is illogical to think that genes are continually producing proteins whether they are needed or not.

Every cell in an animal or plant's body contains the same number of chromosomes according to the species (sex cells, of course, contain half the normal body cell number). Each of these chromosomes carries a collection of genes made from DNA, and it is believed that the DNA is identical in each cell. Yet the cells themselves vary greatly in structure, and above all in function.

After fertilization, for the first few divisions of the new cell, all the resulting cells are identical. Soon, however, something tells the cells to start to become the tissues and organs they are destined to become. By the time the human embryo becomes recognizable as such, many different types of cells have formed. From an original blob of cytoplasm with a nucleus containing both parents' chromosomes arise nervous, muscular, digestive, excretory, reproductive and respiratory organs. Furthermore, each type of cell within an organ uses its genes to a varying extent at different times, depending on its needs at that moment. For example, a liver cell may produce a lot of one enzyme at one time and then change to making another later. A feedback control determines whether a gene is switched on or off. This type of control is self-regulatory and can be compared to a thermostatic control of a central heating system. If we set the thermostat as a certain temperature, the heating system switches itself on as soon as the temperature drops below the predetermined value.

The automatic feedback systems controlling the production of enzymes were first suggested in the early 1960s by Jacques Monod and Germaine Cohen-Bazire of the Pasteur Institute in Paris.

see also...
Genetic code

Gene switches – inducible

In this type of gene regulation the genes that are normally switched off are switched on when required. The control involves an operon, consisting of a number of different genes that lie close to one another:

★ Structural genes. These code for the production of the enzymes involved in a particular set of reactions.
★ Promoter gene. This is the recognition site for RNA polymerase enzyme to bind to.
★ Operator gene. This controls the production of mRNA.

Outside the operon, a regulator gene produced a repressor molecule which can block the operator gene.

Gene induction consists of three steps:

1 Production of the repressor molecule (a protein).
  ★ The regulator gene is on a part of the chromosome distant from the operon and produces a protein called a repressor molecule.
  ★ In the absence of the substrate, the repressor can block the binding site of RNA polymerase.
  ★ This prevents transciption of the genes coding for beta-galactosidase.
2 The inducer (substrate) binds to the repressor protein.
  ★ This is a reversible reaction, which will happen only if the substrate is in high concentration.
  ★ The inducer binds to the repressor, preventing it binding to the RNA polymerase binding site.
  ★ RNA polymerase can then carry out its function and the structural genes can produce the enzyme.
3 Gene transcription and enzyme synthesis.
  ★ Once the repressor protein is deactivated, RNA polymerase can access the operator gene.
  ★ The enzyme is produced.
  ★ This allows the cell to make enzymes only when there is sufficient substrate, i.e. enzyme production is induced by the presence of the substrate.

see also...

Protein synthesis – messenger RNA

Gene switches – repressible

Jacques Monod and Germaine Cohen-Bazire of the Pasteur Institute, Paris, were investigating the enzyme tryptophan synthetase, which is responsible for synthesizing the amino acid tryptophan. They found that when *Escherichia coli* was growing in a medium containing plenty of tryptophan, it stopped making tryptophan synthetase. This type of control, in which the presence of the product helps stop the machinery for making it, is called negative feedback. Thus the amount of tryptophan produced is regulated by the amount already present. The mechanism depends on an operon and maybe summarized in three steps:

1 The repressor is at first inactive.
  ★ When tryptophan is in low concentration the repressor molecule is the wrong shape and cannot bind to the operator site.
  ★ Transcription of the structural genes is not blocked.
2 The repressor is activated.
  ★ When the tryptophan is in high concentrations it binds to the repressor and changes its shape.

3 The repressor binds to the operator.
  ★ The change in shape of the repressor molecule enables it to bind to the operator.
  ★ As a result, transcription is switched off. The structural genes cannot be transcribed because RNA polymerase cannot bind to the operator site.

In gene induction, genes that are induced are normally switched off. The inducer is the substrate.

In gene repression, genes that are repressible are normally switched on. The presence of high levels of the end-product of a metabolic process activated the repressor molecule.

In 1965, Francois Jacob and Jacques Monod received the Nobel Prize for medicine and physiology for their work on gene switches.

see also...

Gene switches – inducible

Gene therapy

The results of genetic screening pinpoint a potential problem, to which gene therapy can sometimes offer a solution. In this technique, medical geneticists aim to alter the genes responsible for certain inherited disorders and insert perfect genes into chromosomes to replace the imperfect ones. The new genes enable the affected cells to function properly. However, this is easier said than done because of the technical problem of getting new genes into the specific cells that need them. There are three main methods for introducing the new gene to chromosomes of recipient cells.

The first method uses viruses to carry genes into cells. One of the first disorders to be treated in this way was adenosine deaminase (ADA) deficiency. ADA is an enzyme vital for making white blood cells in the bone marrow. To treat the deficiency, a sample of bone marrow is taken from a person and the stem cells that normally develop into white blood cells are separated out. Copies of the ADA-producing gene are taken from a human cell and placed inside a virus, which is modified so that it cannot replicate. The modified virus is used to infect the stem cells so that the ADA-producing gene passes into them. Having received the new working genes, the stem cells are returned to the patient where they divide and produce a continuous supply of ADA.

Another method of gene therapy is used to treat cystic fibrosis. A new gene is introduced to the secretory cells of the lungs by inhaling an aerosol spray containing that new gene. The hope is that the genes will pass into the lung cells and switch on the production of the protein needed for the supply of normal mucus, restoring the functioning of the lungs.

A third approach consists of injecting new genes into the bloodstream so that they can reach all the cells of the body. The control units of the genes are programmed to activate the genes only once they are within the target cells. One possible use of this technique is in the treatment of skin cancer (melanoma).

see also...

Cystic fibrosis; Genetic screening

Genetic code

There is a genetic alphabet which is used to translate the linear sequence of bases in DNA into the sequence of amino acids in proteins. The alphabet is A for adenine, U for uracil, C for cytosine, and G for guanine. An alphabet of only four bases is enough to code for the thousands of different proteins that exist in living organisms. Although proteins are complex molecules, they are built from only 20 different amino acids. There is a simple mathematical explanation for this. If each base codes for one amino acid, only four amino acids can be specified. A two-base combination would provide four times as many possible combinations ($4 \times 4 = 16$), but this is still not enough. By 1952, which was quite early in the search for the genetic code, the biochemist Alexander Latham Dounce predicted a three-base combination to code for each amino acid. The prediction was confirmed by George Gamow, a physicist, two years later. It was reasoned that a combination of any three of the four bases would produce 64 possible triplets ($4 \times 4 \times 4 = 64$). However, there are only 20 amino acids, so more than one triplet of bases could code for the same amino acid. Each triplet of bases is called a codon.

Of the 64 codons, 61 code for a specific amino acid. The remaining codons are UAA, UAG, and UGA. These act like full stops at the end of a sentence. In addition, the codon AUG signals the beginning of a message, like a capital letter. The arrangement of codons along the DNA molecule constitutes the genetic code. An average 150 triplets code for one polypeptide and when synthesis of a given protein is necessary the segment of DNA with the appropriate codons is copied in a molecule of mRNA. When the mRNA migrates to the ribosome, its string of codons is paired to anticodons of tRNA, each of which is carrying one of the amino acids necessary to make up the protein.

see also...

DNA structure; Proteins; Protein synthesis – messenger RNA; Protein synthesis – translation

Genetic drift

This is sometimes called the Sewall Wright effect, after the two population geneticists who suggested it. Once Gregor Mendel had shown how genes are inherited and Hardy and Weinberg had demonstrated how these genes are expected to behave in populations, biologists realized that evolution might occur by chance as well as being directed by natural selection. Genetic drift depends on the fact that the fluctuation of allele frequencies in a small population is due *entirely* to chance. If the number of matings is small then the actual numbers of different types of pairing may depart significantly from the number expected on a purely random basis. Genetic drift is one of the factors that can disturb the Hardy–Weinburg equilibrium.

Randomly mating populations are profoundly influenced by the powerful effects of natural selection. In these groups, as adaptive traits are selected others are relentlessly winnowed out as the populations become adapted to the environment by directional selection. Small populations behave differently and are modified by factors different from those operating in larger populations. For example, in small populations there is a change in frequency of genes in a population due to simple chance. Changes of this type are not produced by natural selection. Genetic drift is important to small populations because they would, by definition, have a smaller gene pool (reservoir of genes). This means that any random appearance or disappearance of a gene would have a relatively large impact on gene frequencies. In large populations, any randon change would have little impact because it would be swamped by the sheer numbers of other genes in the population. The bottle-neck effect is an important way that small populations can set the stage for such random events.

By definition, then, genetic drift is the random fluctuation in gene frequencies due to sampling error. In nature, it is important in small populations that become isolated (e.g. island colonizers or koala bears or the giant panda).

see also...

Bottle-neck effect; Hardy–Weinberg equilibrium; Mendelism

Genetic engineering

The science of genetics is undergoing a revolution, largely because of a marriage between genetic engineers and multinational companies with profit in mind. The result is a battery of living factories that can produce proteins on demand. These living factories are yeast, bacteria and other microbes which are modified and harnessed to work for us.

The term genetic engineering is familiar to most people today, having been used in both emotive and trivial contexts by the media and by science fiction writers. Genetic engineering embraces many concepts including gene manipulation, gene cloning, recombinant DNA technology, gene therapy and genetic modification. In a nutshell, genetic engineering means finding specific genes, cutting them out of the chromosomes of one species and splicing them into the chromosomes of another. After the genes have replicated many times, the proteins made by them are harvested. The main advantage of this technique is the mass production of many otherwise scarce and expensive protein-based chemicals which have a direct role in saving lives. For example, the extraction of just 5 mg of the protein somatotrophin (a growth regulating hormone) would require half a million sheep's brains! The same mass of this hormone can be made in one week by 9 litres of genetically engineered bacteria.

The genetic engineer uses some highly specialized equipment and terminology.

★ The source DNA contains the required gene, which is cut out and added to the host DNA.
★ The host DNA is cut to allow the insertion of the source DNA.
★ The recombinant DNA is a hybrid DNA resulting from the fusion of the source DNA and the host DNA.
★ Restriction enzymes are enzymes which cut DNA at specific points, allowing the source DNA to be inserted into the host DNA.

see also...

Cloning genes; Gene therapy; Genetically modified animals; Genetically modified plants; Recombinant DNA

Genetic engineering – a summary

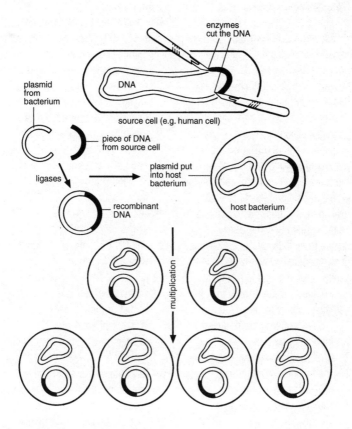

The principles of genetic engineering

Genetic profiling

In the 1980s Professor Alec Jeffreys at Leicester University showed the presence of many variable regions of DNA which did not code for amino acids. These regions were called minisatellites, and there are thousands of them scattered throughout chromosomes, probably having evolved as mistakes during replication of DNA. The number of times that these regions are repeated gives individuality to the profile. If enough regions of DNA are examined, it is possible to obtain a genetic profile which is almost unique to an individual – the change of two people having the same genetic profile is one in a million (unless they are identical twins). A genetic profile is made as follows:

1 Cells with nuclei are obtained from any tissue.
2 DNA is extracted from the nuclei.
3 The DNA is cut into fragments using restriction enzymes.
4 The fragments are separated according to size using electrophoresis. This is the process of subjecting the mixture, on a gel soaked in a conducting solution, to an electric field. Electrically charged DNA fragments move towards oppositely charged electrodes. The positive move to the cathode and the negative to the anode. The rate of movement varies with the size of the fragment.
5 The fragments are transferred to a nylon membrane in a process called Southern blotting. The nylon membrane is sandwiched between the gel and sheets of blotting paper. The DNA fragments move into the membrane.
6 Sodium hydroxide is added to the membrane. It splits the DNA into single strands, leaving the sequence of bases intact.
7 The DNA fragments are identified with a DNA probe. This is a portion of DNA with a base sequence that is complementary to that of the minisatellite. The probe is labelled with a radioactive tracer, which affects X-ray film.
8 X-ray film is placed over the membrane and developed. The result is a pattern of bands like a shop bar code, showing where the probe has bound to the DNA.

see also...

DNA structure; Gene structure; Gene switches; Genetic code

Genetic screening

In 1997 the first recommendation for any nationwide genetic test was made in the United States by the National Institute of Health (NIH). Recommendations of NIH panels are not mandatory but they are generally adopted by the medical profession. This recommendation said that all couples planning to have children should be offered a genetic test for cystic fibrosis (about 800 children are born with cystic fibrosis in the USA each year). With an incidence of one in 30 people carrying the recessive gene, it is the most common inherited disease in the USA. There is a one in 900 chance than an average couple will both carry the mutant gene – and if they do there is a one in four chance that their child will suffer from the disorder.

The information from screening could be used by couples to decide whether to risk having a child in the first place, whether to go ahead but test the fetus with a view to possible termination of the pregnancy if it is afflicted, or whether to complete the pregnancy nevertheless but prepare for the possibility of the child suffering from cystic fibrosis. However, the NIH panels do not suggest routine screening of newborn babies because there is no point in starting treatment before symptoms show, usually at about six months.

The public's general interest in genetic screening is primarily focused on serious genetic disorders within families or among friends. However, there are many other advantages from this research. Medical science has made great progress in understanding the fundamental basis of cancer over the past 20 years, most of which originated from the discovery of cancer-causing genes (oncogenes) and now also anti-oncogenes. Gene research can have an equally important impact on a number of other disorders. For example, the actual cause of Alzheimer's disease is still a mystery, although it is known that some families seem to have a genetic predisposition to it. Recent studies confirm that a gene on chromosome 21 is responsible for a minority of Alzheimer's cases.

see also...

Cancer; Cystic fibrosis

Genetic tracking

enetic tracking is the technique used to find out whether a person has a particular gene in his/her genome. The diagnosis of genetic disorders in unborn children has become a very sensitive issue in modern society. As a result of a technique known as gene tracking, together with chorionic villus sampling, it is now possible to find out how potentially dangerous genes are inherited. Chorionic villus sampling takes a small sample of cells from membranes surrounding the fetus in the womb, and enables DNA from fetal cells to be extracted and tested. Armed with this knowledge prospective parents may choose to continue with an affected pregnancy or terminate it.

In gene tracking, the DNA profile from a particular chromosome of a person who is affected with a disorder is compared with the DNA profiles of the parents. If there are sufficient samples of DNA from a family pedigree, partners can be given information relating to the chances of their child being affected by a harmful gene, before it is conceived.

The process used for the diagnosis relies on the same techniques as genetic profiling. However, the main difference is in the selection of the probe used. A gene probe is just a portion of single-stranded DNA with a radioactive tracer or label which has been cloned to make millions of copies. Very few DNA probes are complementary to the sequence of bases in a functional gene, which actually codes for amino acids. Most probes that are used for genetic tracking recognize a sequence of bases very close to the functional gene, but not the functional genes themselves. These non-functional base sequences tend to be linked to the functional gene and are inherited with them. The closer the marker sequence is to the functional gene, the more useful it is in the diagnosis of a genetic disorder.

see also...

Genetic profiling; Human genome project

Genetically modified (GM) animals

In 1996 Rosie, a cow with a difference, was born. Her birth heralded hope for the survival of thousands of premature babies born each year. Scientists genetically engineered Rosie and eight other cows to produce a human protein in their milk. Early in their fetal development, they were given human genes which make the protein alpha-lactalbumin, normally found in human milk. This protein is a rich and balanced source of amino acids essential for newborn babies. The protein can be produced in bulk in Rosie's milk, purified, and added to powered milk for premature babies. Typically the breed of cow to which Rosie belongs would produce 10,000 litres of milk per year.

Even before the success with Rosie, researchers had created a sheep with human genes as early as 1993. These animals were modified to produce human proteins in their milk to provide a blood-clotting factor needed by haemophiliacs. Alpha-1-antitrypsin is a protein that helps treat cystic fibrosis, and this can also be produced in sheep's milk.

The 1990s saw other life-saving applications of genetically engineered animals, including an insect. For the first time, in 1996, a mosquito that transmits a deadly viral disease was turned into a harmless (if irritating) insect, by modifying its genes. The disease in this case was encephalitis in children, which is relatively rare, but if the same technique could be used against mosquitoes that carry yellow fever and other killer diseases, many thousands of lives could be saved.

For many years there have been attempts at transplanting non-human organs into humans. The poor success rate has been due to rejection of unmatched tissues, which is a problem even between humans. In an effort to overcome this problem, pigs have been developed that contain human genes. Later this century, 'spare part' pigs could be availble with kidneys, hearts and lungs ready to be donated to human recipients.

> ## see also...
> *Cystic fibrosis; Haemophilia*

Genetically modified (GM) plants

In the 1980s, the first commerical crop to be genetically modified to produce its own insecticide was the potato. The insecticide is a toxin produced by the soil bacterium *Bacillus thuringiensis*. It is harmless to humans but kills insect pests, including the Colorado beetle. The toxin breaks down quickly and is harmless to spiders and many beneficial insects. Scientists isolated the gene that controls the production of the toxin and inserted it into a potato plant so that the plant could make the toxin in its leaves. There is concern that harmful insects may evolve resistance to the toxin, and before such a modified plant is used on a large scale, it must receive government approval.

In December 1996, the European Commission approved the sale of genetically modified maize, which contains the genes for herbicide resistance plus a natural insecticide from *B. thuringiensis*. As part of the engineering process a marker gene is inserted; this marks the cells that contain the modified DNA. The marker gives resistance to a commonly used antibiotic. Critics suggest that there is a chance that when cows eat the maize, the gene could pass into bacteria that normally live in the cows' digestive systems and might help to spread antibiotic resistance. Concerns about the antibiotic resistance caused the British government to raise objections to the sale of modified maize in Europe, but the Commission was forced to consider financial implications of closing the European market to the product. The USA export hundreds of million of dollars' worth of maize to Europe each year and to suddenly prevent this could easily start a trade war.

By 1997, governments of those countries which traded in genetically modified products began tightening up their regulations. The Australia and New Zealand Food Authority prepared rules which made it much more difficult to sell food made from genetically modified products. Approval for the sale of such food is subject to meticulous scruitiny. Those foods that are approved must carry a label if they contain more than 5 per cent modified ingredients.

see also...

Genetic engineering

Haemophilia

Haemophilia is an inherited blood disorder in which an essential blood clotting factor is either partly or completely missing. It occurs in all racial groups and affects about one in 10,000 of the male population.

The gene for haemophilia is carried on the **X** sex chromosome. Females are **XX** and males **XY**. A man with haemophilia passes the faulty gene to his daughters, who then carry this gene and may pass the condition onto their sons. A man with haemophilia does not pass the faulty gene onto his sons because they receive a copy of his **Y** chromosome only (their **X** chromosome comes from the mother).

Men are affected because the faulty gene on their **X** chromosome lacks the necessary instructions to produce the clotting factor. Women are usually unaffected because they have two **X** chromosomes and the working copy of the gene tends to override the faulty copy. However, about one-third of female carriers have low clotting factor levels and may be mildly affected. There is a 50 per cent chance of daughters of a carrier mother being carriers themselves. This depends on whether they receive the **X** chromosome containing the faulty gene or whether they receive the **X** chromosome with the normal gene. Sons also have a 50 per cent chance of inheriting the faulty gene or the normal gene. If they inherit the faulty gene, they will always suffer from haemophilia as they do not have a second **X** chromosome to override the faulty copy. About 30 per cent of people with haemophilia have no family history of it. Their affliction is probably due to mutation having taken place in the egg or sperm before fertilization.

Until 1986, some haemophiliacs were treated with contaminated blood products. About 1200 were infected with HIV and over 3000 with hepatitis C. No new infections have occurred since 1986, when heat treatment of blood products reduced the risk of viral contamination. A synthetically produced clotting factor avoids the risk of contamination from viruses found in human blood.

see also...

Chromosomes; Gene mutation; Sex chromosomes; Sex-linked genes

Haemophilia and royalty

A mutation in a sex chromosome of Queen Victoria influenced the fates of both the Russian and Spanish royal families to such an extent that the repercussions echoed around the whole of Europe. The problem began with a mutant allele on Queen Victoria's **X** chromosome which made her a carrier of haemophilia. Two out of the queen's five daughters were also carriers, the princesses Alice and Beatrice. One of her four sons, Prince Leopold, suffered from the condition.

The significance of this pedigree to the Russian royals began with Princess Alice's daughter, Alexandra. She married Tsar Nicholas II and was a carrier of her grandmother's mutant allele for haemophilia. On 12 August 1904, she gave birth to a son, who was named Alexis Tsarevich. He was a haemophiliac but his affliction was considered too serious to be made public and the anguish caused to the royal family clouded objective judgement to such an extent that they employed the infamous Grigori Rasputin to cure the child. Rasputin believed in a religious cult which preached the bizarre dogma of salvation through sin. There is much recorded evidence of his hypnotic powers, which he used to pacify Alexis when he became hysterical during bouts of bleeding, sending him to sleep and lowering his blood pressure. The tsar and tsarina were so obsessed by Rasputin that they granted him powers and protection more suited to royalty. Many state decisions were influenced by Rasputin and it is said that he contributed to the destruction of the credibility of the Russian monarchy, setting the scene for the revolution in 1917 that shaped the destiny of millions.

The same mutated chromosome took part in determining the history of Spain. Queen Victoria's youngest daughter, Beatrice, transmitted the allele to three of her four children, one of whom, Eugenie, became Queen of Spain. Two of Eugenie's sons were afflicted and another was born a deaf mute. It is said that these misfortunes contributed to the mistrust of Eugenie's English background by the true blue-blooded Spaniards.

see also...

Haemophilia

Hardy–Weinberg equilibrium

The concept stems from the Hardy–Weinberg principle. It shows that the frequency of alleles for any character will remain unchanged in a population through any number of generations unless this frequency is altered by some outside influence.

Imagine that **A** is dominant to **a**. A cross between **AA** and **aa** will give an F_1 of **Aa**. If we calculate the gene frequencies in the F_2 generation from a cross between the F_1, we see that one-quarter of the population is **AA**, one-half is **Aa**, and one-quarter is aa. So to determine what fraction of the F_3 generation will be offspring of, say, **AA** and **Aa** crosses, we simply multipy one-quarter by one-half and get one-eighth. In the third generation, then, assuming random mating, we can expect one-eighth of the population to have genotypes resulting from this cross. What part of this generation will be the offspring of **AA** and **AA** crosses? Of **Aa** and **aa** crosses? Each of these combinations can be expected to produce one-eighth of the F_3 generation, so what part of the population will be heterozygous (**Aa**)? By counting the frequency of this type of gamete among the possible combinations, we see that 4 × one-eighth = one-half. And what was the fraction of **Aa** individuals in the F_2 population? One-half! If you work out the results for all possible combinations in the F_3 generation or for any combination in any generation after the F_3 generation, you will see that the ratio of genetic components in a population remains stable. This is why we continue to have blue eyes in our population, even though they are a recessive trait.

The conditions required are as follows:

★ large population
★ random mating – every individual of reproductive age has an equal chance of finding a mate
★ no emigration or immigration – no gene flow
★ no selection pressure – no natural selection
★ no mutation.

Any of the above can occur in natural populations, so the Hardy–Weinberg model is, to a great extent, an artificial one.

see also...

Hardy–Weinberg principle

Hardy–Weinberg principle

In 1908, a lunchtime discussion took place at Cambridge University between the geneticist Reginald Crundell Punnett and his older friend, G.H. Hardy. Punnett said that he had heard an argument that was critical of Mendelian philosophy and to which he did not have an answer. According to the argument, if a gene for short fingers were dominant and the gene for long fingers recessive, then short fingers ought to become more common over each generation. Within a few generations, the critics said, no one in Britain should have long fingers!

Punnett disagreed with the conclusion but he couldn't put forward a good reason. Hardy said he thought the answer was simple enough and jotted down a few equations on a table napkin. He showed the amazed Punnett that, given any particular frequency of genes for normal or short fingers in a population, the relative number of people with short or long fingers ought to stay the same – as long as the population was not subject to natural selection or other outside influences that could lead to changes in gene frequency. (Gene frequency refers to the ratio of different kinds of genes in a population.) Punnett talked the reluctant Hardy into publishing the idea somewhere other than on a table napkin! Hardy thought the idea too trivial to publish. Others were developing the same idea, including a German physician, named Wilhelm Weinberg, so the idea came to be known as the Hardy–Weinberg principle.

The principle was actually developed even earlier by the American, W.E. Castle, but strangely, he never is mentioned in standard text books. The principle is stated today as:

> In the absence of forces that change gene ratios in populations, when random mating is permitted, the frequencies of each allele (as found in the second generation) will tend to remain constant through the following generations.

This lead to the concept of the Hardy–Weinberg equilibrium.

see also...

Hardy–Weinberg equilibrium; Punnet squares and probability

Human chromosomes

Early studies of cells did not reveal much detail because all parts of the cell were about the same optical density. In the late nineteenth century, however, it was found that certain stains could differentiate parts of cells. The visibility of chromosomes was enhanced when cells were stained with basic dyes. In fact, the word chromosome was chosen because some of the basic (aniline-based) dyes were bright colours – hence *chromos* = colour and *soma* = body. Using such dyes on human tissues, Walther Flemming recorded seeing human chromosomes in 1882. It was not until 1923, however, that a close estimate of the number of chromosomes in a human cell was made. Using very thin sections of the tubules from human testes, Theophilis Shickel Painter estimated that there were 48, although he considered the possibility of there being 46.

In 1956, J.H. Tjio and A. Levan made an important discovery that opened the way for much more accurate research on human chromosomes. They devised a method of pressing entire cells flat so that all the chromosomes in a cell were spread and could easily be counted. The complete set of chromosomes could be photographed to illustrate the karyotype of the person. In cells from females, the chromosomes could be sorted into 23 pairs, with the members of each pair being similar in size and structure. In cells of males, however, only 22 matched pairs could be identified, with the two chromosomes of the remaining pair being dissimilar in size and shape. One chromosome of this pair was of medium length, but the other was very short. These are the X chromosome and the Y chromosome, respectively, and are related to sex determination. Female cells have two X chromosomes but no Y chromosome.

The development of tissue culture techniques by Theodore Puck (in collaboration with Tjio) made it possible for karyotypes of white blood cells from blood samples to be used. Such research methods soon showed that many human defects that had puzzled the medical profession for centuries were results of chromosome abnormalities.

see also...

Chromosomes; Sex chromosomes

Human genome project

Since the 1980s advances in the fields of both genetics and medicine have led to developments in medical genetics, which have allowed medical practice to evolve at a rapid pace. The human genome project is an international research effort that aims to analyse the structure of human DNA. Scientists are attempting to map the location of an estimated 100,000 genes. History has shown us that the maps of newly dicovered countries, made by pioneering explorers, were arguably the first steps to the entrepreneurial exploitation of the territories they revealed. As an analogy, we must protect the definitive map of human genes from exploitation by the unethical. It is anticipated that the end product of this research will be the standard reference for biomedical science in the twenty-first century and will help us to understand, and eventually treat, many of the 4000+ recognized human genetic afflictions. The goals to be achieved are:

★ mapping and sequencing of the human genome (all human genes)
★ mapping and sequencing of the genomes of certain other organisms
★ data collection and distribution
★ research training
★ international sharing of ideas in gene technology.

Initial estimates suggested that all this will take up to 15 years. The aim of the enormous research programme is to sequence the four bases making up human DNA (adenine, cytosine, guanine, and thymine). The bases are represented no less than three thousand million times in our genome. If typed in order using their initial letters (A, C, G and T) our sequence of bases would fill the equivalent of 134 complete sets of the *Encyclopaedia Britannica*. The size of an individual gene within the whole length of human DNA is similar in comparison to the size of an ant on Mount Everest! If successful, this mind-boggling project will provide an invaluable reference for medical science in the study of human genetic disorders.

---

see also...

DNA structure; Genetic screening

Huntington's disease

This genetic disorder was introduced to North America in 1630 by two immigrants from Suffolk, England. Its name is derived from the American, George Huntington, who first described it in 1872. He came from a medical family and, as a boy, used to accompany his father on his professional visits. On one such occasion, while driving through a wooded lane in Long Island, he saw two women, mother and daughter, both tall and very thin, bowing and grimacing with uncontrolled movements. This incident affected him so much that he vowed to study the problem when he qualified as a medical doctor many years later. Until that time, the affliction was considered to be the dreadful fate of Providence sent to punish sinners. The original name, Huntington's chorea, came from the Greek word for dancing – a reference to the jerky uncontrolled movements of sufferers.

Huntington's disease is an inherited condition which affects about 1 in 10,000 people in middle age. It attacks nerve cells in the brain and, as these cells slowly degenerate, a person with the disorder loses all control of their mental and physical abilities. The pattern of inheritance is autosomal dominant. Where one parent has the affliction, there is a 50 per cent chance that any children will have it. The problem usually begins with mild symptoms over a number of years. These include loss of memory, clumsiness and depression. As the disorder progresses, physcial and mental control steadily deteriorates. Although people who have inherited the faulty gene can develop symptoms at any age, most only start to show signs between the ages of 30 and 50. This means that some adults only find out that they are at risk of developing the condition after they are parents themselves, by which time they may have passed the gene on to their children. In 1993 the exact position of the faulty gene was discovered on a human chromosome, and this may lead to the possibility of gene therapy being developed for the condition.

see also...

Gene therapy

Independent assortment

This is the law formulated by George Mendel, that genes segregate independently at meiosis so that any one combination of alleles is as likely to appear in the offspring as any other combination.

Chromosomes, not genes, separate and are distributed independently. Stated simply, each of a pair of contrasted characters may combine with each of another pair independently. The logic of this law, sometimes called Mendel's second law of genetics, depends on the following:

★ Factors for different traits are 'sorted' independently of each other – all combinations of alleles are distributed to gametes with equal probability.
★ We now know that, during meiosis, alleles on one pair of homologous chromosomes separate independently from allele pairs on other chromosomes.
★ These factors (alleles) will be inherited in the offspring in predictable ratios determined by the genotype of the parents.

If **Y** = yellow seed (dominant) and **y** = green seed (recessive) and if **R** = round seed (dominant) and **r** = wrinkled seed (recessive), a cross between two genotypes heterozygous for colour and texture would be

RrYy × **RrYy**

The phenotypes (outward appearance) would both be round and yellow. The gametes of each type would be **RY** + **Ry** + **rY** + **ry**.

When they fuse, they would do so to produce the following combination of phenotypes in the following ratio: nine round and yellow: three round and green: three wrinkled and yellow: one wrinkled and green.

It is now known that genes are linked together on chromosomes and so tend to be inherited in groups. The law of independent assortment therefore applies only to genes on different chromosomes.

see also...

Linkage; Meiosis; Mendelism

Inversion mutation

If chromosome segments break away during mitosis or meiosis they may rejoin the chromosome the wrong way around, giving an inversion.

★ The middle piece of the chromosome falls out and rotates through 180° and then rejoins.
★ There is no loss or duplication of genetic material.

As knowledge of the genetic features of *Drosophilia* became more detailed in the 1930s, several obervations were made that were at first inconsistent with what was understood until they were later interpreted in terms of chromosomal abnormalities and confirmed by microscopy. One of these anomalous observations was an inversion. It was discovered that inversions have a profound effect on data relating to gene mapping, which relies on recombination percentages.

As a result of an inversion mutation, there is a problem at meiosis because the two homologous chromosomes cannot align themselves side by side in the same way that they would normally do, because the gene order is reversed in one of them. They overcome this difficulty by forming an inverse pairing loop. If crossing over takes place within the loop, the chromosomes have difficulty in separating and half of the gametes produced are inviable. That is why inversions are important – they prevent recombination taking place between adaptive combinations of linked genes within the inverted region. Crossing over can take place elsewhere within the bivalent, of course. There is no difficulty at meiosis in the inversion homozygote (where the same inversion is present in both homologous chromosomes), and no problem at mitosis in either heterozygotes or homozygotes.

Inversions play an important role in the evolution of many species. Some inversions include the centromere, others do not. Inversions help to keep favourable groups of linked genes together to form one 'supergene'.

see also...

Chromosome maps; Deletion mutation; Duplication mutation; Gene mutation; Linkage; Meiosis; Mitosis; Mutation; Translocation mutation

Klinefelter's syndrome

The discovery of non-disjunction of pairs of chromosomes in sex cells was recognized as being of immense value in providing proof for the chromosome theory of inheritance, but was not at first considered to be of any practical value. Then, this seemingly obscure scrap of knowledge was central in understanding a new discovery in human biology. Techniques were developed in the 1950s for studying human chromosomes, and it became possible to count them accurately. At first it was thought that humans had 48 chromosomes along with the chimpanzee, gorilla and orang-utan, then in 1956, it was confirmed that humans had 46 chromosomes. However, an interesting condition had been recognized for some years by medical practitioners. They thought that certain abnormalities of sexual development were associated with abnormalities in the sex chromosomes. One such condition is called Klinefelter's syndrome.

One in about 500 'male' births produces an individual with a particular set of abnormalities. They have a general male phenotype with the following characteristics:

★ underdeveloped penis and testes, low levels of testosterone and infertility
★ sparse body hair
★ longer limbs than average
★ some characteristically female breast development
★ diminished intelligence
★ high-pitched voice.

Each body cell has 47 chromosomes – 44 + XXY. The XXY individual may arise through fertilization of an XX egg by a Y sperm or through fertilization of an X egg by an XY sperm. Research has shown that non-disjunction of the X chromosomes in ageing oocytes (cells that develop into eggs) occurs more often than XY non-disjunction in sperm formation and so is more important as a determining factor.

As evidence of the nature of the gender-determining mechanism in humans, the important point is that people with at least one Y chromosome have the general phenotype of a male, even in the presence of any number of X chromosomes.

see also...

Chromosomes; Non-disjunction

Lethal alleles

In 1905, the French scientist Lucien Cuenot was carrying out genetic crosses with mice. He crossed a mutant mouse with yellow fur with a pure-breeding (homozygous) normal grey type (**yy**). The result was a 1:1 ratio of yellow to grey mice. He concluded that the yellow mice were heterozygous (**Yy**) and that yellow was dominant. When two yellow offspring were mated together, they produced young in the ratio of two yellow: one grey. This was a strange departure from the expected classical 3:1 Mendelian ratio. Cuenot hypothesized that the allele for yellow must be lethal in the homozygous condition, **YY**. Female

Yy individuals were dissected to test his hypothesis and aborted fetuses were found in them, thus confirming his prediction. The modified two:one ratio was thus explained as shown in the diagram.

Alleles that fail to code for functional proteins in sufficient amounts will be lethal alleles. Dominant lethal alleles are possible but are usually rapidly eliminated because their expression results in death. Exceptions occur when the effects of the lethal allele are not expressed until late in life, e.g. Huntington's disease. Recessive lethal alleles are fatal only in the homozygote.

P
Parental phenotypes:

Parental genotypes:

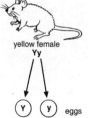
yellow female
Yy

X

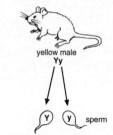
yellow male
Yy

Gametes:

(Y) (y) eggs

(Y) (y) sperm

Possible male gametes

	X	Y	y
Possible female gametes	Y	YY lethal	Yy yellow
	y	Yy yellow	yy grey

Effects of a lethal allele in mice

see also...

Huntington's disease; Mendelism

Linkage

Where independent assortment does not occur, this is often due to genes being linked together on a chromosome and so being passed on to the next generation together.

By the 1920s, scientists knew that the genes of the fruit fly were carried in four groups on four pairs of chromosomes. However, they did not know the positions of these genes in relation to each other. As early as 1911, the possibility had been considered of making maps of chromosomes to show the relative positions of genes on them. The crossing over of parts of chromosomes during meiosis was known as long ago as 1909. Thomas Hunt Morgan suggested that knowledge of crossing over could enable him to deduce the order of genes in each chromosome. His logic was as follows:

Let us hypothesise that genes are arranged in a line along each chromosome, like beads on a necklace. Then during crossing over, a gene may be transferred to another chromosome. If one chromosome carried the genes **ABCDEFGHIJKL** and its partner

had the corresponding recessive genes abcdefghijkl, they could exhange parts so that the resulting chromosomes were **ABCDEFghijkl** and **abcdefGHIJKL**. This process is called recombination, because the genes in a pair of chromosomes are recombined during crossing over.

Now let us suppose that two genes are at opposite ends of a chromosome. When that chromosome breaks during crossing over, both genes may be interchanged, but it is most likely that only one will be interchanged and so they will be separated and appear on different chromosomes. If, however, the genes are close together, separation is less likely. The chance that any two genes will be separated from each other during recombination is thus closely related to their distance apart on the chromosome. If two or more genes are linked together, there is a high chance of them being passed on to the next generation together.

see also...

Chromosome maps; Independent assortment; Meiosis; Mendelism

Meiosis

Meiosis is the process of cell division leading to the production of daughter nuclei with half the genetic complement of the parent cell. Cells formed by meiosis give rise to gametes (sex cells) and fertilization restores the normal cell complement.

As eggs and sperm are both cells, when they combine at fertilization, why don't the resulting offspring have twice as many chromosomes?

As long ago as 1887, the German biologist August Weismann put forward an explanation as to how chromosome numbers might stay constant from one generation to another. He considered two facts:

1 In successive generations, individuals of the same species have the same number of chromosomes.
2 In successive cell divisions, the number of chromosomes remains constant.

To account for these facts, Weismann suggested that the chromosome number is halved. Weismann's prediction was soon verified. Biologists carefully studied the process of sex cell formation in many species and discovered the fundamental truth that halving of the chromosome number takes place in all sexually reproducing organisms.

In 1892 Weismann made another important discovery. He wanted to account for the fact that different offspring from the same parents are not identical. He proposed that inherited characteristics are reshuffled in the formation of sex cells. This would mean that the different sex cells in one organism would contain different combinations of hereditary characteristics (the term 'gene' had not then been invented). He thought that this reshuffling must occur during the halving of the chromosomes. Thus each sperm and each egg is unique. This is the second significant point of meiosis which leads to variation and is a contributory factor for evolution.

see also...

Chromosomes; Fertilization

Meiosis – stages

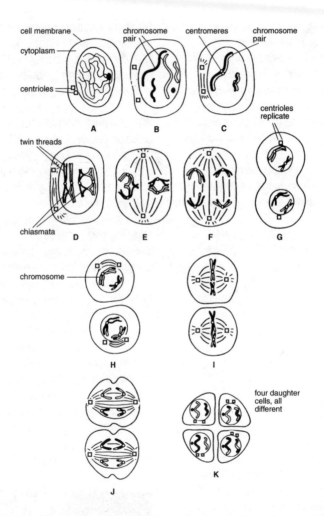

The stages of meiosis, in which a cell divides to form four daughter cells with half the number of chromosomes of the parent cell

Mendel, Gregor

Gregor Mendel was born in 1822 and christened Johann Mendel. His family was of peasant ancestry and lived in Moravia, now a province of the Czech Republic. After simple, early instruction by an uncle, Mendel attracted the attention of the clergy and, at the age of 21, became a monk in the monastery of Brunn. At his ordination he took the name of Gregor, by which he has been known ever since.

Mendel attended the University of Vienna from 1851 to 1853, where he studied mathematics and physics, and then became a teacher of science at the local secondary school. He continued to teach until 1868, when he was made Abbot of the monastery, a position of such responsibility that he did no further teaching and gave up his research. It is interesting to speculate what might have happened had the scientific world of his time responded to Mendel's two publications which led ultimately to the foundation of the science of genetics. Mendel was discouraged by the almost complete indifference of the scientific community to his work. It must, indeed, have been a prime factor in his failure to continue his experiments.

For a period of about eight years, from 1858 to 1866, Mendel conducted his famous breeding experiments with the common garden pea for his long and meticulous studies. His first paper was published in the *Transactions of the Natural History Society of Brunn*, the obscurity of which may account for the indifference with which the paper was received. In 1900, 34 years later, three scientists from different countries independently discovered Mendel's work and realized its significance. They were H. De Vries from the Netherlands, C. Correns from Germany and E. Tschermak from Austria. They completely confirmed Mendel's findings, which were to become the foundation stone of the whole of genetics which has become such a dynamic science in the twenty-first century. Mendel died in 1884 before his work was recognized for what it was. Today, we call him 'the founder of heredity' or even 'the founder of molecular biology' and the first mathematical biologist.

see also...

Mendelism

Mendelism

Mendelism is classical genetics, named after Gregor Mendel. It is the study of inheritance by controlled breeding experiments, first carried out by Mendel in the 1860s. The characteristics studied are controlled by one gene and show a simple dominant or recessive relationship between alleles. Large numbers of progeny from a given cross are counted to find the ratios of various phenotypes (the outward expression of the genes, which may be the outward appearance) and from these the parental genotypes can be assessed. Work of this nature gave the first indication that inheritance is particulate rather than blending.

Mendel formulated two laws to explain the pattern of inheritance he observed in crosses involving the common garden pea, *Pisum sativum*. The first law, the Law of Segregation, states that any character exists as two factors, both of which are found in the body cells (somatic cells) but only one of which is passed on to any one gamete. The second law, the Law of Independent Assortment, states that the distribution of such factors to the gametes is random. Therefore, if a number of pairs of factors is considered, each pair segregates independently.

Mendel gave the name 'germinal units' to those features that he considered to control characteristics in his experimental pea plants. Today these germinal factors are called genes (a term first used by the Danish scientist Wilhelm Johannsen in 1909). The different forms of genes are called alleles, an abbreviation of 'allelomorphic pairs of genes'. It is known that a cell with the full complement of chromosomes for that species (diploid cell) contains two alleles of any particular gene. Each allele is located on one of a pair of homologous chromosomes (chromosomes that pair during meiosis). Only one homologue of each pair is passed on to a gamete (an egg or a sperm). Thus, the Law of Segregation still holds true.

see also...

Gene; Independent assortment; Linkage; Mendel, Gregor

Mitosis

t took 200 years from Robert Hooke's first observations of cells in the 1660s to knowledge of the details of cell division. By 1880, there was general agreement that new cells are formed by equal division, with the nucleus always dividing before the cell. A brilliant contribution to cytology came in 1882 when Walther Flemming published his discoveries. His work was outstanding for a variety of reasons. First, he had access to the most sophisticated microscopes of his day, as German technology in the field of precision optical engineering was second to none; second, new dyes were being developed for staining biological material for microscopic studies. The combination of these two advances enabled Flemming to carry out work that would have been impossible only a generation before.

Flemming never relied on what he saw in dead cells unless he could observe the same in living cells. He also realized that some of the structures that show up clearly in stained dead cells might not be natural – they could be produced by chemical reactions between the dyes and the cellular structues. As a result, his meticulous and careful observations have stood the test of time.

Flemming observed that in the process of dividing, a cell's nucleus passes through an orderly series of changes, which he called mitosis (from *mito* = thread). The events were approximately the same in all the animal and plant cells that he examined. Two daughter cells are produced from a parent cell, and both daughter cells are identical to each other and to the parent cell in all respects. This means that the chromosomes are identical in number and in structure. Flemming observed that mitosis is the method of cell division by which new cells are produced during growth.

Today we can describe mitosis as an ordered process by which the cell nucleus and cytoplasm divide into two. The chromosomes replicate before mitosis and then separate in such a way that each daughter cell inherits a genetic complement, identical to the parent cell.

see also...

Nucleus; Replication

Mitosis – stages

In animal cells, a structure near the nucleus (the centriole) divides (A) and separates (B), giving out radiating protein fibres. The chromosomes take on their characteristic thread-shaped appearance. The centrioles travel to opposite poles of the cell and then become connected by fibres, forming a spindle-shaped arrangement (C). The chromosomes arrange themselves along the equator of the spindle (D), midway between the centrioles. At this stage each chromosome is made up of twin threads joined only at one place, the centromere. After arranging themselves at the equator, the twin threads of chromosomes separate (E). Now one complete set of daughter chromosomes migrates to each pole and, in co-ordination with this, the entire cell divides (F). The daughter chromosomes lose their thread-like apprearance as a new nucleus forms in each daughter cell (G).

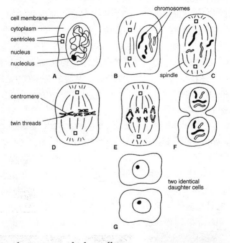

The stages of mitosis, in which a cell divides to form two identical daughter cells, each with the same chromosomes as the parent

see also...

Meiosis; Mitosis; Nucleus

Monohybrid inheritance

This type of inheritance results from crossing parents with one pair of contrasted characters. Gregor Mendel used this method when he formulated his Law of Segregation. He crossed parental pea plants which were pure bred for producing round seeds with pea plants that were pure bred to produce wrinkled seeds. He found that all the hybrids in the F_1 generation produced round seeds, thus showing the dominant character, round, and not the recessive, wrinked. He allowed these to self-pollinate, at random, and found that the missing recessive wrinkled trait reappeared in some of the F_2 plants. Moreover, the ratio of the F_2 plants with the recessive trait to those with the dominant trait was fairly constant, regardless of which contrasted characteristics were involved. The ratio was three dominant:one recessive.

Strangely enough, a detailed statistical analysis of Mendel's results by Ronald Fisher in the 1930s showed that Mendel's figures from his later investigations were, in statistical terms, too good to be true! The implication is that once Mendel had formulated his theory, those results that failed to give the expected ratio were ignored by Mendel or by the gardeners who helped him count his seeds or plants.

The details of Mendel's cross involving one pair of contrasted characters are given below:

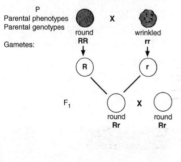

A monohybrid cross using a Punnet square

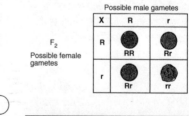

see also...

Mendelism; Punnet squares and probability

Multiple alleles

This means the existence of a series of alleles (three or more) for one gene. Perhaps the most familiar example in humans of incompletely dominant multiple alleles is the ABO blood grouping system. Here three alleles are possible at a single locus (a point on a chromosome where one or more alleles are present). In fact, there are several alleles responsible for blood types but A, B, and O are the most common. People who are homozygous for the A allele produce one kind of red blood cell-surface antigen and are said to be type A. Similarly, those who are homozygous for B antigen are type B. Heterozygotes, AB, carry both alleles, producing both kinds of antigens. They are said to be blood group AB. People who have neither the A nor the B antigen belong to a fourth group – type O. Type O is the most common blood group. It seems that O is a third allele at the same gene locus. It is recessive when paired with either A or B. With three different possible alleles, there are six different possible gentotypes:

Genotype	A antigen	B antigen	Blood type
AA	Present	Absent	A
AO	Present	Absent	A
B	Absent	Present	B
BO	Absent	Present	B
AB	Present	Present	AB
OO	Absent	Absent	O

Although A and B are co-dominant with respect to each other, they are both dominant with respect to the O allele. Note that, although there are three alleles of this gene locus, no one individual can have all three alleles at once. An understanding of the inheritance of blood groups may be used in settling a paternity suit. A woman might allege that a particular man is the father of her child, and testing the blood groups of the man, the woman and her child may help to disprove the allegation. Consider a man with blood group O whose sperm fertilized an egg of a woman with blood Group A: they could not possibly have children with blood groups B or AB. The only possibilities would be children with group A or group O.

> ### see also...
>
> *Co-dominance; Dominance; Mendelism*

Mutagens

Mutagens are changes in the external environment which can have profound effects on mutation rates.

Radiation: Organisms are constantly subjected to various types of radiation. The electromagnetic spectrum extends from long radio waves to extremely short cosmic rays. The amount of energy contained in the radiation becomes larger as the wavelength becomes shorter, and above a certain energy level the rays can penetrate living cells.

Ultraviolet radiation can penetrate cells less well than higher energy rays but is readily absorbed by DNA, causing structural damage at a molecular level. As a result, skin cancer is possible.

Radiation of wavelengths shorter than that found in ultraviolet rays is called ionizing radiation. The energy level is so high that electrons in the irradiated atoms may be knocked out of orbit, thus producing a positively charged ion. Ions, and the molecules which contain them, are chemically much more reactive than the original neutral atoms. The structure of DNA, and hence chromosomes, can be affected by this type of irradiation but any mutations caused will not be maintained in the population unless they occur in organs which produce sex cells.

All organisms are subjected to low levels of ionizing radiation from cosmic rays and from radioactive materials found in the Earth's surface rocks. We are also likely to have additional radiation from human uses of radioactive isotopes – e.g. X-rays and fall-out from nuclear accidents.

Chemicals: Since 1945, a long list of mutagenic chemicals has been compiled and these often have carcinogenic properties. Mutagens have proved valuable in artificially speeding up the production of new variants in horticulture and agriculture. One of the most important of these is colchicine, which interferes with the process of cell division, causing the doubling of chromosomes in cells and producing larger than normal plants.

see also...

Mutation; Polyploidy; Radiation and mutation; Radioactive disasters

Mutation

Mutations are abrupt, heritable changes in single genes or regions of a chromosome. Although the term mutation is now generally used in this sense, historically it was defined more broadly to include alterations in chromosome number and chromosome structure. Mutations constitute the raw material for evolution and are the basis for the variability in a population on which natural (or artificial) selection acts to preserve those combinations of genes best adapted to the environment.

Many mutations may be neutral or 'silent' (i.e. they have no observable effect on the organism). Harmful mutations become evident because they alter the survivial capacity of the organism. Mutations occur randomly and spontaneously and may be induced by environmental factors. Spontaneous mutations arise from errors in replication and different genes may mutate at different rates.

Induced mutations can be due to exposure to environmental factors (mutagens) such as:
★ X-rays: cause breaks in the DNA, leading to chromosomal rearrangements (block mutations) and deletions
★ Ultraviolet rays: cause point mutations, including base-pair substitutions, insertions and deletions
★ Chemical agents: these include:
 – base analogues, which mimic and substitute for normal bases in DNA synthesis, leading to mispairing
 – reactive chemicals, which add chemical groups to or delete them from normal bases, leading to mispairing during DNA replication (examples are benzene, formalin and carbon tetrachloride)

Genes mutate at known rates, but the rate varies depending on the gene involved. Most mutations occur in somatic (body) cells and are not inherited. Those that occur in sex cells may be inherited.

see also...

Chromosomes; Deletion mutation; Duplication mutation; Gene mutation; Inversion mutation; Mutagens; Replication; Translocation mutation

Neurofibromatosis

Neurofibromatosis (Nf) is a genetic disorder which causes tumours to form on nerve tissue anywhere in the body. There are two types, Nf1 and Nf2. The pattern of inheritance for both types is autosomal dominant. Each child of an affected parent has a 50 per cent (one in two) chance of inheriting the faulty gene and of having the disease. However, 50 per cent of people with Nf have no family history of the condition. They are probably affected because of a spontaneous genetic mutation which took place in the egg or sperm before fertilization.

Early signs of Nf1 include six or more coffee-coloured marks on the skin before a child is five years old, freckling in the armpits and lumps on or just below the surface of the skin. Large tumours can form on the skin. These tumours are not usually harmful in themselves but can severely affect a person's appearance, particularly if they are on the face. Tumours pressing on the nerves in the eye, the ear or internally on the spine can affect sight, hearing, speech, and even lead to paralysis and early death. A person with Nf1 may have learning difficulties and an increased risk of epilepsy. Some bones may be affected – the spine may be curved and the long bones below the knee and below the elbow may be malformed. Nf1 affects about one in every 2500 people worldwide: there are more than 23,000 people with Nf1 in the UK alone.

Nf2 is less common, affecting one in every 35,000 people worldwide. It nearly always involves hearing loss due to tumours on nerves in both ears. Diagnosis can be difficult, but the effects of the tumours in the body are always serious and may be life-threatening.

Living with Nf can be hard for people whose appearance is severely affected. Tumours on the skin tend to start to appear around puberty and may increase in number throughout life, so some adults will have many swellings anywhere on their bodies. Some tumours can be removed surgically.

see also...
Gene mutation

Non-disjunction

In 1916, Calvin B. Bridges published a paper called *Non-disjunction as proof of the Chromosome Theory of Heredity*. In this, and in later studies, he provided evidence that genes are physically associated with chromosomes. He studied the way in which a gene for bright red eyes, called vermilion, was transmitted in *Drosophila*. Vermilion eyes are caused by a recessive allele which, like the allele for white eyes, seemed associated with the X chromosome. A cross between vermilion-eyed females and the normal red-eyed males would usually give rise to normal red-eyed females and vermilion-eyed males. Very occasionally, no more than one in 2000 flies, the cross produced a vermilion-eyed female! If the theory of recessive alleles were correct, this female must have inherited two recessive alleles for vermilion, yet she should have inherited the allele for red eyes from her normal red-eyed father. Where did the two recessive genes come from?

Bridges deduced that the female parent's pair of X chromosomes failed to separate during egg formation. Non-disjunction is the term used to describe this failure of the chromosome pair to separate.

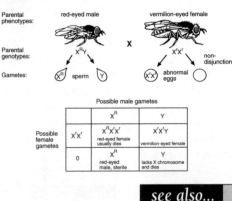

Non-disjunction in fruit flies

see also...

Dominance; Sex-linked genes

Nucleic acids

Nucleic acids are organic acids whose molecules consist of chains of alternating sugar and phosphate units with nitrogenous bases attached to the sugar units (nucleotides). They occur in the cells of all organisms.

In 1868, a young Swiss biochemist named Friedrich Miescher discovered a chemical which proved to be the most significant substance in the structure of a gene. He named the chemical nuclein because it existed only in the nuclei of cells. He needed lots of cells for his investigations and the source of his research material was pus from discarded surgical bandages obtained from a local hospital. The pus provided a plentiful supply of white blood cells from which he could obtain the nuclei.

Meischer eventually changed his original bizzare supply of material to another source. Each spring he visited the Rhine falls above his home town of Basel at the time when salmon were leaping the falls in large numbers. With the help of local fisherman, he obtained enough nuclear material in the form of salmon sperm to last him through the following year. Meischer soaked the sperm in strong salt solution and precipitated the strands of nucleic acid by adding water. This had to be carried out at low temperatures, so in those days before refrigerators the only way was to work in an unheated room in the middle of winter.

Working under these rigorous conditions, Meischer succeeded in analysing nucleic acid to find the chemical elements present. Besides phosphorus, he discovered the presence of carbon, hydrogen, oxygen and nitrogen.

Other scientists continued the research into nucleic acid. The German, Albrecht Kossel, discovered that it contained combinations of atoms built around two basic carbon–nitrogen configurations called purine and pyrimidine rings. Kossel isolated two pyrimidines, which he named cytosine and thymine, and two purines, which he named guanine and adenine.

see also...

DNA structure; Protein synthesis – messemger RNA; Nucleus; Protein synthesis

Nucleus

The nucleus was first described in 1700 by Antoni van Leeuwenhoek, a Dutch draper who spent his spare time grinding lenses for the microscopes that he made. He placed some fish blood on a clean piece of glass and, through his microscope, saw little oval particles – red blood cells. He noticed that in some of these particles there was 'a clear sort of light in the middle'. Without knowing it, he had observed the nucleus of a living cell. Ironically, if he had examined the red cells of human blood or the blood of any other mammal, he would not have seen a nucleus because mammalian red blood cells are among the very few cells that do not have nuclei.

In 1781, Felix Fontana, found oval bodies inside the skin cells of an eel. Once again the nucleus was described, but its importance was not recognized. It was the Scottish botanist, Robert Brown, who first established in 1883 the idea that a nucleus is a normal part of a living cell by examining hundreds of cells of different plants. Although better known to chemistry students for his description of the motion of particles (Brownian motion), Brown was the first to recognize the cellular nucleus as the central part of a cell.

Observations of animal cells were more difficult than observations of plant cells because of their greater variety of shapes and sizes, but many microscopists of the late nineteenth century subscribed to the idea that animal cells also have nuclei. Today, it is recognized that a nucleus is the cellular organelle that controls all the activities taking place in the cell. It is found in all living cells except mammalian red blood cells and in phloem sieve tubes (cells which carry sugar around plants). It is bounded by a nuclear membrane, perforated by nuclear pores that allow exchange of material with the cytoplasm. The nucleus contains genetic material, which arranges itself into chromosomes during cell division.

see also...
Chromosomes

Pesticide resistance

With all of human technology, we have not managed to eliminate even one species of unwanted insect or other pest. The gene pool in species of insects with such large and prolific populations is such that there will always be the odd mutant which has a gene for controlling the production of an antidote to a pesticide. This principle also holds for pesticide resistance in all other species, for example, rats, as long as their gene pool is large enough.

Resistance arises as a result of selection acting upon natural variation and is an example of evolution in action. There are several sources of evidence demonstrating this but some of the most enlightening date from the 1960s. At that time geneticists were carrying out research with the fruit fly, *Drosphilia*, to show how natural resistance to DDT could be produced by selective breeding. The life cycle of this species is about three weeks, and after several generations artificial selection produced completely resistant populations. Similarly, after several generations, mutations for resistance to other insecticides have been artificially induced in *Drosophila*

by irradiation. Doubtless such mutations occur naturally in wild populations from time to time, although at much lower frequencies.

Some pesticides act as gene switches in insects, switching on genes to control the production of certain enzymes, which can break the pesticides down to harmless components. Resistance may be dominant, recessive, or co-dominant. It often appears to be under the control of a single gene. This is the case with the insecticide dieldrin; however, at least three different interacting genes cause resistance to DDT in the housefly.

Although some species evolve resistance rapidly, others seem to lack the potential altogether despite many years of exposure. However, once resistance arises it can spread rapidly, as evidenced by the malaria-carrying mosquito in India.

see also...
Enzymes; Gene switches; Mutagens

Phenylketonuria

Phenylketonuria (PKU) is caused by the inherited failure of the body to produce an enzyme which normally breaks down the amino acid phenylalanine to another amino acid, tyrosine. It is caused by an autosomal recessive gene.

This conversion normally takes place in the liver, and in its absence the level of phenylalanine in the blood rises. PKU was discovered in 1934 by the Norwegian physician A. Folling, who demonstrated that abnormal metabolites (phenylketones) were present in the urine of affected children. A fetus, homozygous for the recessive allele for PKU, is normal before birth because the excess phenylalanine is removed through the placenta and converted in the liver of the mother. Within three days of birth, however, the phenylalanine level of the blood increases above normal. Within a few weeks it will become as high as 60 mg or more per 100 cm^3 of blood (a normal concentration is only 2–3 mg per 100 cm^3). The levels remain high in the blood and interfere with brain development, leading to learning difficulties. The muscles and cartilages of the legs may be defective, so many individuals with untreated PKU cannot walk properly. Sufferers are also likely to have fairer than normal skin and hair because of a deficiency of tyrosine, which is needed for melanin formation.

Control of PKU is difficult because phenylalanine is present in most proteins in a normal diet. It is one of the essential amino acids because we cannot synthesize it. An artificial diet containing a mixture of synthetic amino acids, except phenylalanine, is given to an infant who suffers from the condition soon after birth. The affected baby can develop relatively normally, although some phenylalanine is needed to build body proteins. This can be obtained from fruits and vegetables, which are allowed along with the mixed amnio acids. The object is to supply just enough phenylalanine for the body's requirements with little or no excess to accumulate in the blood. In the UK one in 15,000 babies are born with PKU.

Pleiotropy

This is a situation in which one gene is involved in the production of several seemingly unrelated characteristics in the phenoype of an organism. For example, in a study of laboratory rats a whole range of abnormalities, including an enlarged heart, closed nostrils, tracheal obstruction and lack of elasticity in the lungs, were all found to be caused by a single mutant allele. The normal form of this gene was later shown to produce a protein essential for the formation of cartilage. Such widespread damage arises because cartilage is a vital part of many body organs. In view of the complex pathways of normal metabolism, it is not surprising that effects like this are extremely common – so much so that most genes are now thought to be pleiotropic.

One example of pleiotropic inheritance in humans that is similar to the example described in rats is Marfan syndrome. Anyone who receives the dominant gene for Marfan syndrome will express a number of traits that may seem unrelated, but all result from one basic defect – the production of an abnormal form of connective tissue which includes bone and cartilage. Three categories of defects result:

1 Skeletal defects. Here the long bones, fingers and toes grow longer and there is also an unequal growth of the bones of the rib cage, giving rise to a protrusion (pigeon chest) or a depression.
2 Cardiovascular defects. This results in a weakness in the walls of the aorta and a weakening of the heart valves.
3 Eye defects. The lens may be displaced.

Another example found in humans is sickle-cell anaemia. It is one in which the gene that codes for the production of the haemoglobin molecule undergoes a point mutation to produce a genetic disorder called sickle-cell anaemia. This results in multiple effects on the phenotype, mainly due to impaired circulation. The effects include brain damage, heart defects, kidney defects, enlargement of the spleen and skin lesions.

see also...

Dominance; Mutation; Sickle-cell anaemia

Polyploidy

Polyploidy is the condition in which a cell or organism contains three or more times the haploid number of chromosomes. Polyploidy is far more common in plants than in animals and very high chromosome numbers may be found: for example, octoploids and decaploids (containing eight and ten times the haploid chromosome number). Polyploids are often larger and more vigorous than their diploid counterparts and the phenomenon is therefore exploited in plant breeding, in which the chemical colchicine can be used to induce polyploidy. Polyploids may contain multiples of the chromosomes of one species or combine the chromosomes of two or more species. Polyploidy is rare in animals because the sex-determining mechanism is disrupted. For example, a tetraploid XXXX would be sterile.

Complete polyploid humans are, as might be expected, quite rare and the few cases known are either spontaneously aborted fetuses or stillborn. A few live for a matter of hours. Gross and multiple malformations occur in all cases, and these reflect the extreme genetic imbalance of the individuals concerned. D.H. Carr described the chromosome complement of 227 aborted fetuses in the 1960s. Of these, chromosome abnormailities were found in 50: two were triploids and one was tetraploid. As a result of similar research, it is now estimated that about 15 per cent of all spontaneously aborted fetuses are either triploids or tetraploids. The orgin of polyploid embryos in humans is difficult to explain. Although the fusion of a normal haploid gamete with a diploid gamete is possible and would produce a triploid individual, there is very little evidence of this taking place in mammals.

Polyploidy has been important in the evolution of some species. Hybridization between two different species sometimes produces a completely new species. C.L. Huskins in 1930 and C.J. Marchant in 1963 described a classic case of this type of evolution in the saltmarsh grass, *Spartina.*

see also...

Autopolyploidy; Allopolyploidy; Human chromosomes; Mutagens; Non-disjunction

Population genetics

To a geneticist the term 'population' has a precise meaning. A population is a local community of a sexually reproducing species in which the individuals share a common gene pool. Geneticists generally restrict themselves to populations of sexually reproducing organisms in which there is random mating, with each member having a equal chance of mating with any other member of the population. Because the sharing of genes takes place essentially by Mendelian inheritance, such local communities are also referred to as Mendelian populations.

The largest exclusive group of potentially interbreeding individuals which can comprise a Mendelian population is a species, but it is rare for an entire species to form one random mating group. What we generally find is that the species is made up of a large number of local populations with varying degrees of gene flow between them. At one extreme there may be an open population which is subject to immigration of genes from other inter-communicating groups within the species, and at the other end of the range a closed population with the only source of new alleles being mutation. Population genetics is a study of how genes are distributed and inherited within populations of various types, and of the frequencies of these genes in populations.

In a Mendelian population the sum total of the genes within the reproductive cells of all its members constitutes the gene pool. If reproduction in a population takes place randomly, with each individual having an equal chance of mating with every other member of the population, we can think of all the genes in the gametes as belonging to one large pool of genes. In thinking about the inheritance of genes in populations, it is the gene pool, and the way in which its composition may or may not change over a number of generations, which is of interest to the population geneticist. It is the sample of the gene pool, passed on from one generation to the next, that results in evolution.

> ### see also...
> *Genetic drift; Hardy–Weinberg equilibrium; Mendelism; Mutation*

Porphyria

Porphyria is an inherited disorder which took a part in shaping the history of Britain and, arguably, the USA. It was reputed to the cause of the 'madness' of King George III. His illness precipitated the notorious Regency Crisis, in which the Prime Minister of the day, William Pitt, was in danger of being expelled. It was popularly believed that, during one of his bouts of illness, the king made decisions which sowed the seeds of the American War of Independence and thus contributed to the loss of the American colonies. One of the symptoms of porphyria was observed by George III's physicians, although its significance was not realized. This was the dark red colour of the king's urine. The colour was due to a red chemical called porphyrin, part of haemoglobin of the blood. Normally porphyrin is broken down by an enzyme, so its presence in the urine is a sign that this critical enzyme is missing, the result of a mutant gene.

The clinical seriousness of a similar defect was first brought to the attention of the medical world in 1908 by the London physician Sir Archibald Garrod. He was studying a rare disorder called alcaptonuria, in which the urine turns black and smelly after eating certain foods. He found that the smelly substance was a chemical resulting from the failure to break down a certain substance in the food. Garrod predicted that the symptoms of alcaptonuria arose because an enzyme was not working properly.

The more serious problem of porphyria affects the nervous system. Attacks begin in the nerves and eventually reach the brain, resulting in paralysis, delirium and agonizing pain. The cause is a mutant autosomal recessive gene which is responsible for producing the porphyrin-breaking enzyme.

The British royal connection with the disorder can be traced back to Mary Queen of Scots in the sixteenth century. In 1786, the disease claimed Frederick the Great of Prussia. Edward, Duke of Kent, the father of Queen Victoria, suffered severely from the condition and died of it in 1820.

see also...

Enzymes

Position effect

The effectiveness of a gene in expressing itself depends on its relation to its neighbouring genes. This is called the position effect, and was studied in detail by H.J. Muller. He suggested that the effects of a gene might partly depend on overlapping regions between neighbouring genes. The exact limits of an individual gene would therefore depend upon its neighbour and would constantly shift at each recombination during meiosis. Muller and his colleague Daniel Raffel hypothesized in 1940 that the gene might be divided into still smaller pieces which under some conditions could act as independent units; for example, in recombination.

The hypothesis was followed up by Guido Pontecorvo who, in 1952, while looking for subdivisions of the gene, pointed out that mutations at several sites close together might produce different effects. This depends on whether the sites were all on one partner of a chromosome pair or whether they were distributed between both partners. He used the fungus *Aspergillus nidulans* for his research. This fungus can synthesize a chemical called biotin, but its ability to do so can be destroyed by a recessive mutation at any one of three closely grouped sites on the chromosome. If there were two mutations on the same chromosome of a chromosome pair, their effect would be masked by the other normal chromosome carrying the dominant factor. On the other hand, if there was a mutation at one site on one member of a chromosome pair and another at another site on its partner, the fungus would die because it could not produce biotin.

The interpretation of these observations pointed to one of two possibilities: either three separate genes were involved or there were three mutant sites within one gene. Pontecorvo considered the second possibility to be the more likely, which meant that the gene could be separated into constituent parts. This pioneering work has led to the human genome project.

see also...

Human genome project; Meiosis; Mutation

Proteins

Proteins are complex nitrogen-containing molecules found in a whole range of structures in our bodies. Hair, muscle, silk and tendon have structures made up of protein. Less visible forms of protein regulate our biochemistry and therefore determine whether we live or die. Haemoglobin carries oxygen in the blood; hormones carry chemical messages around the body; and (most important of all) biological catalysts called enzymes are essential for the series of controlled chemical reactions that we call life.

The name, protein, was coined in 1838 by Gerardus Johannes Mulder. He realized that protein was the most important organic chemical yet discovered in the body, and gave it a name that meant 'primary' or 'holding first place'.

Besides being responsible for the basic chemistry of life processes, proteins also give us individuality – every organism forms its own kinds of protein. As with fingerprints, no two individuals have identical sets of proteins, because their synthesis is under the control of unique combinations of genes.

The structure of proteins was first studied in detail by one of the most outstanding chemists of all time, the German Emil Fischer. As long ago as 1819, a French chemist, M.H. Braconnet, had isolated amino acids, the building bricks of proteins. Fischer showed how these units were linked together in chains joined by bonds called peptide linkages. Chains of amino acids joined in this way are called polypeptides (= many peptides). It is known today that there are 20 amino acids, making up all the known proteins by being arranged in various combinations. The number of ways that 20 units can be combined with repeated sequences is astronomical. A medium-sized protein is made up of about 5000 atoms in total. These giant molecules were so difficult to analyze that progress was not really made until the 1940s. Federick Sanger spent ten years working out the amino acid sequence of insulin, a relatively small protein with 51 amino acids. He was awarded a Nobel Prize for this work.

see also...

Enzymes; Protein synthesis

Protein synthesis

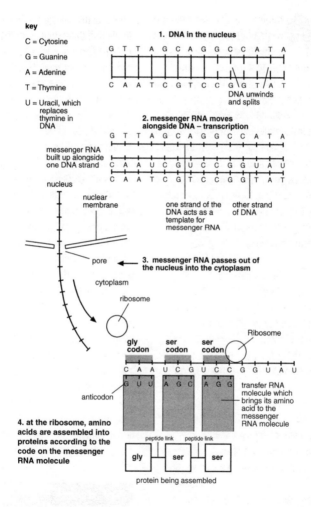

key

C = Cytosine

G = Guanine

A = Adenine

T = Thymine

U = Uracil, which replaces thymine in DNA

1. DNA in the nucleus

| G | T | T | A | G | C | A | G | G | C | C | A | T | A |

| C | A | A | T | C | G | T | C | C | G | G | T | A | T |

DNA unwinds and splits

2. messenger RNA moves alongside DNA – transcription

| G | T | T | A | G | C | A | G | G | C | C | A | T | A |

messenger RNA built up alongside one DNA strand

| C | A | A | U | C | G | U | C | C | G | G | U | A | U |

| C | A | A | T | C | G | T | C | C | G | G | T | A | T |

one strand of the DNA acts as a template for messenger RNA

other strand of DNA

nucleus

nuclear membrane

pore ◄— **3. messenger RNA passes out of the nucleus into the cytoplasm**

cytoplasm

ribosome

Ribosome

| **gly codon** | | | **ser codon** | | | **ser codon** | | |

| C | A | A | U | C | G | U | C | C | G | G | U | A | U |

| G | U | U | A | G | C | A | G | G |

anticodon

transfer RNA molecule which brings its amino acid to the messenger RNA molecule

4. at the ribosome, amino acids are assembled into proteins according to the code on the messenger RNA molecule

peptide link   peptide link

| gly | — | ser | — | ser |

protein being assembled

A summary of protein synthesis

Protein synthesis – messenger RNA

Although the discovery of transfer RNA (tRNA) was essential to the understanding of protein synthesis, it did not explain the role of DNA in the process. Scientists reasoned that some complementary pattern or template must exist to ensure that the RNA molecules brought their amino acids into line in the correct sequence for making a particular protein. In the early 1960s, a brilliant suggestion welded together all the previous observations in this field and made a framework for an explanation. The stroke of genius came from Francois Jacob and Jacques Monod at the Pasteur Institute, Paris. They boldly proposed that a newly discovered RNA molecule with a DNA-like structure had the function of carrying genetic instructions from the nucleus to the ribosomes. They named this special type of RNA 'messenger RNA' (mRNA). The hypothesis offered a working model for the vital transmission of information from DNA to the protein assembly area.

More researchers directed their efforts towards the idea and soon a picture emerged in which RNA was being continually sent out from the genes to the ribosomes, where it was used and then replaced by fresh RNA. The genetic message was thought to be transferred from DNA to RNA by a base-pairing mechanism. The first step must be for the two spirals of the DNA molecule to separate, exposing the bases. A complementary strand of messenger RNA could be built up alongside one of the DNA strands in a process called transcription. In 1963 RNA was proved to be a single-stranded molecule transcribed from only one strand in the DNA molecule. An enzyme called RNA polymerase splits the two spirals of DNA, like someone undoing a zip fastener, thus allowing the RNA to build up along the exposed bases in one chain. In this base pairing, a new base, uracil, substitutes for thymine. In this way the genetic message from DNA is used as a code for the synthesis of RNA, which itself codes for the amino acids to be brought to the ribosomes for protein synthesis.

see also...

DNA structure

Protein synthesis – transcription

It was not until the end of the 1930s that scientists realized that ribonucleic acid (RNA) was present in both animal and plant cells. Before this, it was thought that DNA was a component of animal cells and that RNA was to be found only in plant cells. The Belgian, Jean Brachet, and the Swede, Torbjorn Caspersson, were the first to show the universal presence of RNA in living cells in the 1940s. They also found that most of the RNA occurs outside the nucleus, although some is contained in special parts of the nucleus called nucleoli. This discovery guided them into thinking that RNA might form some sort of link between the nucleus and the rest of the cell.

Caspersson and Brachet continued their research and discovered that cells that specialize in making lots of protein are those with most RNA. For example, cells from the silk glands of the silk moth that make silk, which is pure protein, have an abundance of RNA. It could be possible that the synthesis of protein depended on the presence of RNA.

By 1943, techniques had been discovered to separate all components of the cell. Albert Claude of the Rockefeller Institute managed to produce a sediment of the smallest cell components, the ribosomes, which are visible only under the electron microscope. Fine chemical analysis showed that the ribosomes were a mixture of RNA and protein, and that the RNA in these particles accounted for most of that found in the whole cell.

In 1950 work began to identify the site in the cell where protein was made. After exposing cells to amino acids labelled with radioactive carbon, researchers found that ribosomes had more than twice as much radioactivity per gram of protein as other cell components. This indicated that the ribosomes were sites for the assembly of protein from their amino acid building blocks. In the mid 1950s it was shown that an enzyme that broke down RNA prevented protein synthesis in the cell, but when more RNA was added, protein synthesis started again.

Protein synthesis – translation

After it was found that RNA at the site of the ribosomes was essential for making proteins, researchers concentrated their minds on how RNA carried out its function. By 1957 20 different kinds of soluble RNA had been identified, each linked to one kind of amino acid. It was shown that the soluble RNA, or transfer RNA (tRNA) as it was called, brought amino acids to the ribosome and lined them up in the correct sequence to form a particular protein molecule. From physical and chemical studies it is now known that tRNA contains only a single nucleotide chain of about 80 nucleotides. The complete nucleotide sequence of a tRNA molecule was first elucidated in 1965 by a team of scientists at Cornell University under the direction of Robert W. Holley.

Several modifications to the model have been made since then, but the essential role of tRNA in protein synthesis remains that it combines with specific amino acids to bring them in a line at a ribosome. Chains of amino acids, linked together by peptide links eventually build up to form proteins, which may be structural proteins or enzymes. An enzyme formed in this way will control one of thousands of biochemical reactions that go on in the body at every instant, and will determined the very nature of a particular cell.

The role of tRNA cannot be separated from the role of mRNA in protein synthesis: tRNA carries the amino acids to the ribosomes and arranges them along the mRNA molecule. For each amino acid there is a different tRNA molecule with a specific triplet of unpaired bases called an anticodon. This pairs with the corresponding codon in the mRNA in the ribosome. Thus the amino acids are aligned according to the base sequence in the mRNA in a process called translation.

see also...

Protein synthesis

Punnet squares and probability

Punnet square is a grid, names after its inventor, R.C. Punnet, in 1905. You can use such grids to predict the results of genetic crosses such as Mendel's.

The alleles that could be present in the female gametes are placed on the left of the grid, and the alleles that could be present in the male appear on the top of the grid (these could be reversed). The alleles from both are combined in the relevant squares of the grid. This shows all the possible different pairings of the alleles and hence all the different possible genotypes of the offspring.

Besides showing all the possible allele pairings, a Punnet square also gives the probability for each pairing. That is, it shows how often on average a given pairing is likely to occur. For example, the grid made form a cross between two genotypes, Rr × Rr shows a probability ratio of one RR:two R:one rr. If there were, say, 100 offspring from this cross, you might expect 25 RR, 50 Rr, and 25 rr. However, a small number of offspring is very unlikely to show the exact ratio of offspring predicted by the square. The larger the number of crosses involved, the closer the results will be to the Punnet square probability ratio.

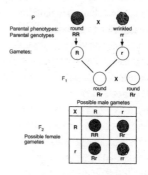

A monohybrid cross using a Punnet square

You can see this effect by flipping two coins. In each flip, the coins may land two heads, two tails, or one head and one tail. The probability of one head and one tail is twice as great as that of either two heads or two tails. By chance, a quite different ratio may appear in a small number of flips. However, the more times you flip the coins, the closer the results will be to the probability ratio.

see also...

Mendelism

Radiation and mutation

All living things are constantly exposed to a certain amount of radiation. The sources of this background radiation include radioactive materials in the Earth's crust and cosmic radiation from outer space. This background radiation is in part responsible for the mutations that occur in all organisms. Radiation is a series of sub-atomic particles or high-energy rays. When radiation hits an atom it can strip off an electron and release a large amount of energy, so any biological molecule hit by radiation may be destroyed. This in itself is enough to cause considerable damage to a living cell.

It has been known for a long time that an increase in radiation levels causes an increase in the mutation rate – the rate at which genes are altered. This knowledge has been used in genetic experiments to produce mutations in organisms used for research into genetics (radiation has produced some useful plant varieties). It has also been known that even very low levels of radiation have an effect on the mutation rate. These findings have real significance for us in the modern nuclear and technological age.

The human race has increased radiation levels in several ways: we use X-rays for medical purposes, we test nuclear weapons, and the dumping of radioactive waste from nuclear power stations is a problem which has not been fully solved with modern technology. This increase in radiation may be only a fraction of the naturally occuring background radiation, but it must be emphasized that any increase in radiation levels will increase the mutation rate.

Radiation has its greatest effect on the bone marrow and liver cells, which are very active tissues in the body, and also on the sex organs where it may cause mutations which will be passed on to the next generation.

Radioactive waste is sometimes disposed of on the sea bed in the deeper oceans or under the surface of the land. Whichever method is used, there will always be routes by which radiation may return to the environment if containment fails.

see also...

Mutagens; Mutation; Radioactive disasters

Radioactive disasters

The legacy of some nuclear disasters has had lasting effects on our environment. In the early morning of 10 May 1977, a massive explosion shook the Dounreay nuclear plant on the coast of northern Scotland. Two kilograms of a potassium and sodium mixture had been deposited in a shaft containing radioactive waste in which there was sea water. The subsequent uncontrolled release of energy was so violent that a huge concrete lid was blown off and its steel top-plate was thrown 12 m to one side. Five-tonne concrete blocks at the mouth of the shaft were badly damaged and scaffolding 40 m away was distorted. An eyewitness described a plume of white smoke blowing out to sea.

By 1987, investigations had begun into a cluster of chidlhood leukaemia cases around Dounreay. A link between radioactive particles and the number of cases of leukaemia was suspected, but it was not until 1995 that it was revealed that the ground around the shaft had been littered with radioactive particles.

In 1986, there was enormous media coverage of the Chernobyl incident. At 9.00 a.m. on 28 April 1986, the first signs of higher than usual radiation levels were monitored at a nuclear plant in Stockholm, Sweden. Winds had been blowing from the south into Scandanavia for several days, carrying with them radioactive materials from the nuclear accident that had occured at the Chernobyl nuclear power plant, 130 km north of Kiev. The plant had experienced a meltdown. This occurs when the cooling system fails and the radioactive core overheats to melting point. The molten radioactive material may burn through the ground and contaminate the water table.

By March 1989, the effects of this disaster were becoming obvious in farms just outside the 30 km exclusion zone around Chernobyl. Deformed animals were born in alarming numbers: some lacked heads, limbs, eyes or ribs; most pigs had deformed skulls. Radiation levels on the farms were 148 times as high as background level. Women from the area were advised not to conceive and the incidence of cancer cases doubled.

Recombinant DNA

Recombinant DNA is a hybrid DNA molecule resulting from the fusion of the source DNA and the host DNA in the process of genetic engineering. The bacterium, *Escherichia coli* normally lives in the human intestine as a harmless microbe. It is one of the most common bacteria used for gene cloning and, like many other bacteria, it has a large circular chromosome and a small circle of DNA called a plasmid. Plasmid rings can be cut open with the same restriction enzymes used to isolate the fragment of source DNA. When the source fragments and the cut plasmids are mixed together, the appropriate 'sticky ends' join up in the presence of a binding enzyme, a type of ligase. The result of this splicing is recombinant DNA, a plasmid containing the source fragment, which must be inserted into the host bacterium. The bacteria are mixed with calcium salts which make their cell walls permeable to the altered plasmids. Once inside the host cell, the new plasmids replicate. Their new DNA sequence is also reproduced each time the host cell reproduces. In this way recombinant DNA is cloned. The host bacteria are allowed to reproduce in ideal conditions and when there are sufficient numbers of bacteria they are broken open and the protein harvested.

After splicing and cloning the recombinant DNA, the genetic engineer has a population of bacteria, some of which contain plasmids and some which do not. The problem is to find the bacteria that contain plasmids and separate them from the rest. This is not as difficult as it might seem because most plasmids used in cloning give antibiotic resistance to their hosts. The host population is subjected to antibiotics and those that die are the ones without the plasmids. The rest are cloned.

see also...

Cloning genes; Gene splicing

Replication

One important implication of Watson and Crick's discovery of the double helical structure of DNA was that it suggested a mechanism for self-duplication. Watson and Crick clearly recognized this potential copying mechanism and this has become one of the most famous quotations in molecular biology:

> It has not escaped out notice that the specific pairing we have postulated immediately suggests a possible copying mechanism for the genetic material. The pairing rule of bases in the DNA molecule ensures that one nucleotide sequence of one chain in the double helix must have a fixed relation to the other chain. Therefore, if the sequence in one part of the fixed chain is ATAGGC, the sequence along the second complementary chain must be TATCCG, where A = adenine; T = thymine; G = guanine; and C = cytosine. If the two nucleotide strands separate, each will serve as a template to govern the structure of the other.

With this in mind, Crick suggested that the two spirals in the DNA molecule might begin to unwind and separate in the presence of a supply of new nucleotides. The new nucleotides would attach themselves by their bases to the correseponding partners on the template. Each newly paired nucleotide chain would grow until a complete complement to the first was formed. Eventually two DNA molecules would exist where only one existed before.

One difficulty in accepting this idea was the need for the whole DNA molecule to unwind into its halves before duplication could begin. As Crick pointed out, a single chromosome must be equivalent to around ten million turns of the spiral. The idea of all this unwinding before duplication begins is hard to believe. Crick predicted that the synthesis of two new chains might begin as soon as the original ones had started to unwind.

see also...

DNA double helix; DNA structure; Watson and Crick

Selective breeding in animals

For thousands of years the selected pedigree animal has been prized. Such animals have a recorded ancestry for many generations. The dairyman wants selected pedigree cattle from which to breed milking cows; the hunter wants a pedigree dog. Selective breeding aims to produce animals progressively better suited to human needs. Until relatively recent times, this form of breeding by artificial selection was somewhat hit-or-miss: before the work of Gregor Mendel in the 1960s, no one had a clear idea of even the most basic principles of heredity, so the development of a new strain of plant or a new breed of animal was by trial and error. Ever since humans began to domesticate animals hundreds of different varieties of animal breeds have gradually been developed. All modern chickens have been selectively bred from the jungle fowl of the Far East; the ancestor of the pig is the wild boar; that of the cow, the wild ox; and that of the overweight Christmas turkey, the wild and slender flying turkey hunted by the original pilgrims of New England.

An example of the value of selective breeding concerns the cattle of India. For centuries, the typical Indian cow has been an unkempt and semi-wild creature. Breeding was haphazard and milk production pitifully small. The cow is considered sacred by the Hindus and roams the village streets unmolested and unattended, picking up food as it goes along. On government farms an attempt was made to improve these animals. The work began by giving them better food, but this did not significantly improve their milk yield. The problem was that the cows, being nervous creatures, were easily frightened and did not relax enough to yield very much milk when milked by hand. Eventually after several years of selective breeding, a breed of less excitable cattle was produced, which allowed people to hand milk them. After 20 years, milk production was increased threefold. Today, a knowledge of selective breeding is combined with artificial insemination to increase the production of desirable breeds.

Selective breeding in plants

One of the first pioneers of selective breeding in plants was the American Luther Burbank. He was responsible for developing a huge range of improved varieties of plants. Perhaps his most famous contribution to plant breeding was his success with potato crops. One day in 1871, while examining a field of potatoes in Massachusetts, Burbank noticed fruit growing on one of the plants. Although potato plants normally have flowers, they seldom produce fruit. New plants tend to be grown from potato stem tubers rather than from seeds. He saved the seeds and planted them. He checked the potatoes growing on the resulting plants and saw that they differed from plant to plant. Some were large, others were small; some produced many potatoes, others produced few. One plant had more potatoes that were also larger and smoother than the others. He selected this plant for future breeding by asexual means. It was named the Burbank variety of potato, and soon became popular throughout the USA.

Burbank's success was an example of mass selection, where one plant is chosen for breeding from a larger number of individuals. Mass selection is probably the oldest type of artificial selection. The first people to cultivate plants always saved the seeds for the next planting from plants that produced the best yield. The offspring are then most likely to have the desired traits and, over countless generations, modern crops have been evolved from their wild ancestors. Cereals, for instance, have been developed from wild grasses. Cauliflowers, broccoli, cabbage and brussels sprouts all belong to a family which originated as wild plants near sea shores.

Mass selection can also produce strains of disease-resistant plants. For example, suppose a fungus has spread throughout a wheat-produing area. Almost all the wheat has been killed except for two or three plants which survived. Seeds from these healthy plants are grown in the following year. Again the fungus attacks the crop, but this time more plants survive. Over several years this cycle is repeated and each year more plants survive to produce a fungus-resistant strain of wheat.

Semi-conservative replication

In 1958 Matthew Stanley Meselson and Franklin William Stahl confirmed the Watson–Crick hypothesis. They grew the bacterium *Escherichia coli* in a culture solution containing glucose, mineral salts without nitrogen, and ammonium chloride in which all the nitrogen was 'heavy nitrogen' (^{15}N), instead of the normal nitrogen (^{14}N). After allowing the bacteria to divide 14 times, Meselson and Stahl had obtained bacteria with heavy nitrogen in their DNA. They then changed the medium by adding an excess of ordinary ammonium chloride containing ^{14}N, so that in future growth the DNA formed by the bacteria would have ^{14}N in it.

If the Watson-Crick model were correct, all bacteria formed in the first cell division after the addition of ordinary ammonium chloride would have DNA molecules, half containing the original heavy nitrogen and half containing the new ordinary nitrogen. At the second cell division, half of the offspring would contain some heavy nitrogen while half would have only ordinary nitrogen. To find out whether this had indeed happened, they had to check the weight of the DNA at successive divisions of the bacteria to see if the DNA molecules contained heavy nitrogen. They found the density of DNA extracted from the bacteria before and after addition of ordinary ammonium chloride to the culture medium.

Their first sample came from bacteria before the addition of ordinary ammonium chloride. It was composed of only heavy-nitrogen DNA. The next sample came from bacteria which had divided once after the addition of ordinary ammonium chloride. This DNA was less dense. The third sample came after two bacterial divisions and contained two densities of DNA – one the same as the second sample and one of ordinary density. Finally, after four divisions, there was a trace of the intermediate-density DNA, but most of the DNA was of ordinary density. The intermediate-density DNA was made of one strand containing heavy nitrogen and one containing ordinary nitrogen. The method of copying had conserved half of the original DNA molecules.

see also...

Replication

Sex chromosomes

These chromosomes carry the gender-determining genes. As the sex chromosomes are usually different in structure and genetic make-up in the two sexes, there are modified patterns of inheritance of the genes that are located on them. Indeed, the association of modified patterns of inheritance with a particular pair of chromosomes within the complement provided some of the earliest and most convincing evidence for the chromosome theory of heredity.

There are two types: the X chromosome and the Y chromosome. In the heterogametic sex (XY) they can usually be distinguished from the other chromosomes, because the Y chromosome is much shorter than the X chromosome with which it is paired (unlike the remaining chromosomes, the autosomes, which are in similar homologous pairs).

In species having almost equal numbers of males and females, gender determination is genetic. Very occasionally, as in the mosquito, a single pair of alleles determines gender but usually whole chromosomes are responsible. The 1:1 ratio of males to females is obtained by crossing the homogametic sex (XX) with the heterogametic sex (XY). In most animals, including humans, the female is XX and the male is XY, but in birds, butterflies and some fish, this situation is reversed. In some species gender is determined more by the presence of two X chromosomes than by the presence of the Y chromosome, but in humans the Y chromosome is important in determining maleness. Rarely, gender is subject to environmental control, in which unequal numbers of males and females develop. Female bees develop from fertilized eggs and are diploid (with their full chromosome complement), while males develop from unfertilized eggs and are therefore haploid (with half the normal chromosome complement). The numbers of each gender are controlled by the queen bee, which can lay either unfertilized eggs, with the haploid number of 16 chromosomes, or fertilized ones with the diploid number of 32 chromosomes.

see also...
Linkage; Sex-linked genes

Sex-linked genes

The first person to put forward the idea of sex-linked genes was Thomas Hunt Morgan, as long ago as 1909. For his research into this field he was eventually awarded the Nobel Prize for medicine and physiology in 1933, when he was a professor at Columbia University. Morgan chose for his studies the fruit fly, *Drosophila melanogaster*, which means the 'black-bellied honey lover'. This fly is about two millimetres long and appears in vast numbers, particularly on over-ripe bananas left lying around in the summer. It can complete its life cycle in a couple of weeks and so can demonstrate hereditary changes quickly and in large numbers in successive generations. Many flies can be kept in a relatively small space and they can be fed on a cheap synthetic diet.

Morgan had been impressed by the work of the Dutch botanist Hugo de Vries, who had been researching into mutations in plants. Morgan began to try to induce mutations in *Drosophilia* using X-rays and other mutagens. After a year of experimentation while he bred vast numbers of fruit flies, he at last recognized the kind of mutation that he had hoped for: of all the thousands of normal red-eyed flies, there was a male fly with white eyes. Morgan bred the single white-eyed male with several of its red-eyed sisters. He obtained 1237 offspring, all with red eyes. This was not surprising because he had assumed that red was dominant to white.

When he crossed the red-eyed F_1 generation together, expecting a ratio of 3:1 red-eyed to white-eyed, he was amazed to find 3470 red-eyed flies and 782 white-eyed flies, and every white-eyed fly was male! In later generations, Morgan found white-eyed female mutants and so he was able to switch the cross to a white-eyed female with a red-eyed male. All the daughters were red-eyed and all the males were white eyed. His conclusion was that the gene for white eyes was linked to the X sex chromsome.

see also...

Dominance; Mutagens; Sex chromosomes

Sickle-cell anaemia

Sickle-cell anaemia is an inherited blood disorder in which there is a defect in the structure of haemoglobin, which carries oxygen in red blood cells. Haemoglobin picks up oxygen from the air in the lungs and carries it around the body to where it is needed. A person with sickle-cell anaemia has sickle haemoglobin in the blood (so called because the red blood cells change from the usual disc shape, with bowl-like depressions, to crescent sickle-shaped cells). As the red blood cells in people with the sickle-cell condition do not last as long as ordinary red blood cells, sufferers may be anaemic from time to time.

The pattern of inheritance for sickle-cell anaemia was discovered in 1949, when James Van Gundia Neel at the University of Michigan showed that it is caused by an autosomal recessive gene. A person who inherits one faulty gene for sickle-cell will be a carrier. Carriers are usually unaffected but can pass the faulty gene on to any children they may have. If one or both parents is a carrier, there is a 50 per cent chance that each child of theirs will be a carrier. Carriers of sickle-cell are sometimes said to have 'sickle-cell trait'. A child who inherits two copies of the faulty gene (one from each parent) will have sickle-cell anaemia. If both parents are carriers, there is a 25 per cent chance of this happening.

Sickle-cell anaemia is most common in people of African or African-Caribbean origin, but may also occur in people of Mediterranean, Middle Eastern and Asian descent. This is probably because being a carrier of sickle-cell gives some protection against malaria. One out of every 300–400 Black Britons is born with sickle-cell anaemia and about one in every 8–10 Black Britons is a carrier of the trait.

In the 1950s, Linus Pauling analysed normal haemoglobin and sickle-cell haemoglobin in detail and found that, of some 300 amino acids making up their molecules, only one was different in the sickle-cell type!

Tay Sachs disease

Tay Sachs disease is an inherited metabolic disorder caused by the absence of an enzyme called hexosaminidase A (hex-A). Without this enzyme, a fatty substance found in the nervous system, GM(2) ganglioside, builds up in the brain cells and destroys them. Children who suffer from the condition rarely live beyond the age of five years. The condition is an example of a late-acting lethal gene.

Tay Sachs disease is named after a British doctor called Warren Tay and an American neurologist called Bernard Sachs. Tay, in 1881, described eye changes in a person with the disease and Sachs recognized that inheritable changes took place in cells of affected children.

The pattern of inheritance for Tay Sachs is autosomal recessive. A person who inherits one faulty gene will be a carrier. Carriers are usually unaffected but can pass on the faulty gene to any children they might have. If one or both parents is a carrier, there is a 50% chance that each child of theirs will also be a carrier. A child who inherits two copies of the faulty gene (one from each parent) will have Tay Sachs disease. If both parents are carriers, there is a 25% chance of this happening.

The disease occurs most often in people of Central and Eastern European (Ashkenazi) Jewish descent. Approximately one in 25 Ashkenazi Jews are unaffected carriers, compared with one in 250 of the general population. However, the disease also occurs among other groups, for example non-Jewish French Canadians.

Babies with Tay Sachs may appear to develop normally until about six months and then they become hypersensitive to noise and their eyesight begins to fail. At about one year, convulsions and fits occur, their muscles become progressively weaker and their co-ordination deteriorates. Swallowing becomes difficult and the lungs do not function properly. Tragically children with Tay Sachs disease die by the age of five because their nervous system has been destroyed.

see also...
Lethal alleles

Thalassaemia

Thalassaemia is an inherited blood disorder in which there is a defect in the structure of haemoglobin. Haemoglobin is a protein which is contained in red blood cells. It picks up oxygen from the air in the lungs and carries it to where it is needed in the body. A person who does not have enough haemoglobin is anaemic. There are several types of thalassaemia but beta thalassaemia is most common.

The pattern of inheritance is autosomal recessive. A person who inherits one faulty gene will be a carrier. Carriers are usually unaffected but can pass on the faulty gene to any children they may have. If one or both parents is a carrier, there is a 50 per cent chance that each child of theirs will also be a carrier. Carriers of thalassaemia are sometimes said to have thalassaemia minor or beta thalassaemia trait. A child who inherits two copies of the faulty gene (one from each parent) will suffer from thalassaemia. If both parents are carriers, there is a 25 per cent chance of this happening.

Beta thalassaemia affects mainly people of Mediterranean, Middle Eastern, or Asian origin. It is thought that thalassaemia is more common in these parts of the world because peopl who carry the thalassaemia gene are protected from the more severe forms of malaria. Among people of Southern Italian or Sicilian ancestry living in New York, beta thalassaemia occurs in about one birth in 2400. Thalassaemia is rare in Northern Europeans.

There are about 200,000 people in Britain who are carriers of beta thalassaemia and about 600 who are homozygous for the gene. Children who are born with the condition appear healthy at birth but become anaemic between the ages of three and 18. Once diagnosed, sufferers have regular blood transfusions every four to six weeks for the rest of their lives. Most children who have these tranfusions live normally into their teens but to extend their lives they also need treatment with an iron-reducing drug, desferal, to avoid damage to their heart, liver and other major organs.

Transforming agent

In 1928 Frederick Griffith led the way to the discovery of how chemicals could carry information between cells. He was studying some of the closely related bacteria that cause pneumonia, the bacteria form mound-shaped colonies with either rough or shiny surfaces. The rough (R) type is harmless but the smooth (S) type can cause the disease and is therefore called virulent. Griffith injected the S type into mice and transferred the blood from the infected mice from one mouse to another. By doing this, he eventually obtained and R type. He found that about one bacterium in every ten million mutates to give rise to the R type. Griffith heated the virulent S type to 60°C and injected them into the mice. They proved to be harmless. He then injected the mice with a mixture of heat-killed virulent S bacteria and live non-virulent R bacteria. Some of the mice died of the bacteria infection. On examining the blood of the dead mice, he found live virulent S bacteria identical to the virulent heat-killed bacteria that had been originally injected. He deduced that some factor must have passed from the heat-killed virulent S bacteria to the living non-virulent R bacteria, causing them to change their nature. This was the transforming agent.

In the 1940s, a team led by the Oswald Theodore Avery in the Rockefeller Institute indentified the transforming agent as DNA. They selectively attacked the known chemical components of the bacterial cells with enzymes, one by one. When the carbohydrate coat of the S type was broken down, the transforming agent remained intact. They then attacked the protein of the cells with protein-splitting enzymes. Again the transforming properties were unaffected. Only the nucleic acids, DNA and RNA, were left. When DNA was broken down with an enzyme, the transforming agent failed to be effective. They published their results in 1944:

> The active transforming material … contains no demonstrable protein, unbound lipid, or serologically reactive polysaccharide, and consists principally, if not solely, of a highly polymerised viscous form of deoxyribonucleic acid.

Translocation mutation

If chromosome segments break away during mitosis or meiosis they may rejoin a different part of the same chromosome or another chromosome, giving a translocation.

★ Translocation involves the movement of a group of genes between different chromosomes.
★ A piece of one chromosome breaks off and joins on to another chromosome.
★ The consequence is a chromosome deficient in genes.

As with inversions, translocations do not affect mitosis in any way, but they cause complications at meiosis as heterozygotes. The reason is that two pairs of chromosomes now have some homologous parts in common. There may be infertility due to the production of inviable gametes, and the interchanges are important because they link together genes in different pairs of chromosomes and so prevent independent segregation. As with inversions, they regulate the pattern of genetic variation in natural populations.

In 1959 it was shown that translocation can occur between human chromosomes nine and 22. The tips of the two chromosomes are exchanged resulting in chronic myeloid leukaemia (CML). The disorder is characterized by an excess of some types of leukocytes. The excess of leukocytes is associated with a reduction of red blood cells and severe anaemia. The abnormal chromosome is found only in the myeloid cells of the bone marrow, which produce the leukocytes. CML is actually a form of cancer of these cells. It may be that these cells are particularly susceptible to agents that cause breakage at this point, or the breaks may occur in all cells but survive and divide only in these cells of the bone marrow. It could also be that leukaemia develops first and causes the breaks. Because translocations often result in infertile individuals (due to failure of normal gamete formation), they can be included and used in insect pest control.

see also...

Deletion mutation; Duplication mutation; Gene mutation; Independent assortment; Inversion mutation; Linkage; Meiosis; Mitosis; Mutation

Turner's syndrome

As a result of non-disjunction there are some people with 45 chromosomes instead of 46 (the principle of this occurrence is the same as the one that is responsible for Klinefelter's syndrome). People with Turner's Syndrome have only one sex chromosome, X, and have the genotype XO with regard to sex chromosomes. Their total chromosome complement is 44 + XO.

Individuals with Turner's syndrome can be produced as a result of non-disjunction during sperm cell formation, producing sperm devoid of a sex chromosome. On fertilizing a normal egg, this would give an XO embryo. Non-disjunction during egg cell formation can result in an egg without any sex chromosomes, and on fertilization with a normal Y-bearing sperm, this would give a YO embryo. However, it would be impossible for a YO individual to survive, as there are too many essential genes on the X chromosome, and YO embryos are aborted. Cells from people with Turner's syndrome have no Barr bodies while those from people with Klinefelter's syndrome do. (Barr bodies are named after Murray Barr, who first described them in nerve cells in 1949. They appear as tiny specks in the nuclei of cells.) Also, white blood cells of affected women have a small drumstick-like structure attached to the nucleus. Both Barr bodies and drumsticks are thought to represent non-functional X chromosomes in females. Cells from affected males lack both Barr bodies and drumsticks.

At an Olympic Games in the 1980s a European woman who showed remarkable strength in field events was found to be genetically abnormal, lacking Barr bodies and drumsticks in her cell nuclei. Her strength was believed to be due to male characteristics. These days, sex tests involving Barr bodies are compulsory for certain athletics competitions.

People with Turner's syndrome have an overall female appearance.

see also...

Klinefelter's syndrome; Non-disjunction; Sex chromosomes

Watson and Crick

In 1953, two scientists produced a hypothesis which resulted in one of the most famous discoveries of all time. The British researcher Francis Harry Compton Crick collaborated with the young American James Dewey Watson in a partnership which proved to be perhaps the most rewarding example of Anglo-American co-operation in the history of science. Watson is quoted as saying 'He taught me physics, I taught him biology. It worked out rather well.'

Crick had spent the Second World War designing mines for the Royal Navy and had then changed the direction of his interests, becoming fascinated by molecular biology. He worked with X-ray diffraction techniques on large protein molecules in America but then moved to Cambridge, where he met Watson, who had a research fellowship in biology. A brilliant, unpredictable young man, Watson was a child prodigy, studying at a college in America at the age of 15. This 'dynamic duo' of physicist and biologist set about building models of DNA until they made one which fitted the known facts. They meticulously constructed exact scale models with precise lengths and angles to match the lengths and angles of the chemical bonds in the molecule.

Amazingly, the whole model was worked out within six weeks. It is said that during the work Watson continued his habits of having a lesiurely café breakfast each morning and a game of tennis each afternoon. The term 'double helix' is used to describe the final shape of two chains of nucleotides twined around each other like a rope ladder. The work was published in 1953 in one of the most famous papers ever to have appeared in the scientific journal *Nature*.

There was immediate world-wide appreciation of the sheer elegance and simplicity of Crick and Watson's work. They shared the Nobel Prize for medicine and physiology, which they received in 1962 with Maurice H.F. Wilkins. Crick built a symbolic copper helix outside his house in the centre of Cambridge.

see also...

DNA double helix; DNA structure

Further Reading

Bainbridge, B.W. *Genetics of Microbes* (Blackie, 1980).
Brown, T.A. *Genetics, a Molecular Approach* (Van Nostrand Reinhold, 1989).
Brown, T.A. *Gene Cloning, An Introduction* (Chapman & Hall, 1990).
Carter, Michael *Genetics and Evolution* (Hodder & Stoughton, 1992).
Crick, F.H.C. The Genetic Code: *Scientific American*, **207** (4), 66–74, 1962).
Dyer, A.F. *Investigating Chromosomes* (Edward Arnold, 1979).
Ford, E.B. *Understanding Genetics* (Faber, 1979).
Jenkins, Morton *Inheritance and Selection* (Simon and Schuster, 1992).
Jenkins, Morton *Teach Yourself Genetics* (Hodder & Stoughton, 1998).
Lewis, K.R. and John, B. *The Matter of Mendelian* Heredity (Longman, 1972).
Mazia, D. The Cell Cycle: *Scientific American*, **230** (1), 54–69, 1974.
Mendel, G. *Experiments in Plant Hybridisation* (Oliver and Boyd, 1965)
Peters, J.A. *Classic Papers in Genetics* (Prentice Hall, 1959).
Suzuki, D.T., Griffiths, A.J.F. *et al.*, *An Introduction to Genetic Analysis* (W.H. Freeman and Co., 1989).
Watson, J.D. *The Double Helix* (Penguin Books, 1970).
Williams, J.G. and Patient, R.K. *Genetic Engineering* (IRL Press, 1988).

Also available in the series